Verena Nitsch

Haptic Human-Machine Interaction in Teleoperation Systems

Verena Nitsch

Haptic Human-Machine Interaction in Teleoperation Systems

Implications for the Design and Effective Use of Haptic Interfaces

Südwestdeutscher Verlag für Hochschulschriften

Impressum/Imprint (nur für Deutschland/only for Germany)
Bibliografische Information der Deutschen Nationalbibliothek: Die Deutsche Nationalbibliothek verzeichnet diese Publikation in der Deutschen Nationalbibliografie; detaillierte bibliografische Daten sind im Internet über http://dnb.d-nb.de abrufbar.
Alle in diesem Buch genannten Marken und Produktnamen unterliegen warenzeichen-, marken- oder patentrechtlichem Schutz bzw. sind Warenzeichen oder eingetragene Warenzeichen der jeweiligen Inhaber. Die Wiedergabe von Marken, Produktnamen, Gebrauchsnamen, Handelsnamen, Warenbezeichnungen u.s.w. in diesem Werk berechtigt auch ohne besondere Kennzeichnung nicht zu der Annahme, dass solche Namen im Sinne der Warenzeichen- und Markenschutzgesetzgebung als frei zu betrachten wären und daher von jedermann benutzt werden dürften.

Coverbild: www.ingimage.com

Verlag: Südwestdeutscher Verlag für Hochschulschriften GmbH & Co. KG
Heinrich-Böcking-Str. 6-8, 66121 Saarbrücken, Deutschland
Telefon +49 681 37 20 271-1, Telefax +49 681 37 20 271-0
Email: info@svh-verlag.de

Zugl.: München, Universität der Bundeswehr München, Diss., 2012

Herstellung in Deutschland (siehe letzte Seite)
ISBN: 978-3-8381-3268-6

Imprint (only for USA, GB)
Bibliographic information published by the Deutsche Nationalbibliothek: The Deutsche Nationalbibliothek lists this publication in the Deutsche Nationalbibliografie; detailed bibliographic data are available in the Internet at http://dnb.d-nb.de.
Any brand names and product names mentioned in this book are subject to trademark, brand or patent protection and are trademarks or registered trademarks of their respective holders. The use of brand names, product names, common names, trade names, product descriptions etc. even without a particular marking in this works is in no way to be construed to mean that such names may be regarded as unrestricted in respect of trademark and brand protection legislation and could thus be used by anyone.

Cover image: www.ingimage.com

Publisher: Südwestdeutscher Verlag für Hochschulschriften GmbH & Co. KG
Heinrich-Böcking-Str. 6-8, 66121 Saarbrücken, Germany
Phone +49 681 37 20 271-1, Fax +49 681 37 20 271-0
Email: info@svh-verlag.de

Printed in the U.S.A.
Printed in the U.K. by (see last page)
ISBN: 978-3-8381-3268-6

Copyright © 2012 by the author and Südwestdeutscher Verlag für Hochschulschriften GmbH & Co. KG and licensors
All rights reserved. Saarbrücken 2012

Abstract

Teleoperation systems are designed to allow their user(s) to perform manipulation tasks in a remote environment, usually in real-time. Application fields range from teleoperated micro-assembly, over minimally-invasive surgery, to large-scale assembly of heavy parts. Human work performance with modern teleoperation systems is hampered by two issues in particular: poor movement accuracy and excessive force application, both of which frequently result in handling errors, damage to equipment and slow work performance. Haptic interfaces, which transmit tactile information to the user of these systems, promise to address these problems. Is it sensible to exploit the sense of touch in the design of human-machine interfaces? This question seems certainly justified, considering that the development and construction of haptic interfaces usually translate into additional financial expenses, which cannot necessarily be recovered through higher retail or service prices. For industries employing teleoperated systems, it is therefore paramount to establish (1) whether haptic interfaces yield measurable advantages over conventional visual interfaces and, if they do, (2) how they are effectively designed and employed to optimise work performance. Yet, systematic, scientific investigations regarding tangible benefits of haptic interfaces for teleoperation systems are scarce, as current efforts are constrained by an atheoretical, fragmented and biased evidence base. As a result, transparent guidelines, which could advise system developers on the best use of haptic signals in teleoperation systems, are still lacking.

The present work closely examines the principles of human-machine interaction with haptic interfaces, discusses previous research efforts and presents new evidence. Furthermore, a synthesis of empirical evidence is presented from which transparent guidelines for the design and effective use of haptic interfaces are derived. These are combined with previously established guidelines on the design of multi-modal interfaces in form of an interactive software package, the "HMI (Human-Machine Interface) Design Guide". It is hoped that accessible and transparent interface design guidelines will encourage system developers to take more notice of basic principles of haptic human-machine interaction during the design and development process and thus promote interdisciplinary work in this field.

Acknowledgements

I wish to express my sincere thanks to the people whose support made this thesis possible.

First and foremost, I would like to express my gratitude to my supervisor, Prof. Dr. Berthold Färber, on whom I could always count for his support and helpful advice. Thanks are also extended to Profs.-Ing. Gunther Reinhart and Kristin Paetzold, who kindly agreed to support my endeavour in interdisciplinary research. I would also like to gratefully acknowledge the enthusiastic support of Prof. Dr. Michael Popp, whose door was always open to me when I was in need of inspiration and motivation. Further thanks are extended to all of my present and former colleagues at the Human Factors Institute of the Universität der Bundeswehr München, but in particular to Dipl.-Psych. Ines Karl and Dipl.-Ing. Guy Berg, who were always both willing and able to support me and who somehow managed to make this whole process fun.

I would also like to express my appreciation to the many engineers and computer scientists at the Technische Universität München and the German Aerospace Centre (DLR) in Oberpfaffenhofen, who collaborated with me in numerous interesting projects as part of the Collaborative Research Centre on "High Fidelity Telepresence and Teleaction" (SFB453). I am particularly grateful for the professionalism and kind support of my collaborating partners at the Institute for Machine Tools and Industrial Management (*iwb*) at the TU München, whose work forms a central part of this thesis. Thank you all for keeping an open mind and for your inexhaustible patience with me and my psychologist's way of thinking.

This thesis would not exist without my family and my friends. Thank you for your never-ending support and for sharing so many laughs with me.

Last, but by no means least, I would like to thank all the individuals who took part in the research. Thank you for allowing this research to take place. Your interest in this subject and my studies kept me motivated.

LIST OF CONTENTS

FIGURES 5
TABLES.. 7
CHAPTER I. AN INTRODUCTION TO TELEOPERATION SYSTEMS AND HAPTIC
INTERFACES 9
   1.   WHAT IS A TELEOPERATION SYSTEM? 9
      1.1.   General set-up of a teleoperation system 9
      1.2.   Definition of terms 10
   2.   APPLICATIONS OF TELEOPERATION SYSTEMS 11
      2.1.   Benefits of the industrial employment of teleoperation systems 12
      2.2.   Current challenges in the industrial employment of teleoperation systems 12
   3.   THE HUMAN-MACHINE INTERFACE 15
      3.1.   Multi-modal interfaces 16
      3.2.   Haptic human-machine interfaces 17
          3.2.1.   What exactly is "haptics"? 17
          3.2.2.   The physiological basis of haptic perception 18
      3.3.   The role of haptic feedback in human-machine interaction.................. 19
      3.4.   Haptic human-machine interfaces: an overview 21
          3.4.1.   The origins of haptic human-machine interfaces for teleoperation systems.... 21
          3.4.2.   Classifications of haptic interfaces 22
      3.5.   Current approaches to haptic interface research and development.......... 27
          3.5.1.   System-centred research and development 28
          3.5.2.   User-centred research and development 29
          3.5.3.   Application-centred research and development 30
   4.   MOTIVATION OF THE PRESENT WORK 33
CHAPTER II. THE EFFECT OF HAPTIC FEEDBACK FROM THE REMOTE ENVIRONMENT
ON TASK PERFORMANCE... 35
   1.   AN INVESTIGATION OF VIBRATORY SIGNALS AND THEIR EFFECTS ON TASK PERFORMANCE
DURING TELEOPERATED MICRO-ASSEMBLY 37
      1.1.   The effect of directional vibrotactile feedback on force application and performance
speed 37
      1.2.   The effect of non-directional vibrotactile feedback on force application and
performance speed 38
      1.3.   The effect of non-directional vibrotactile feedback on performance error and task
completion times 39
      1.4.   Research aim 41
      1.5.   Research hypotheses 41
      1.6.   Method 42
          1.6.1.   Participants... 42
          1.6.2.   Experimental apparatus.. 42
          1.6.3.   Experimental design... 44
          1.6.4.   Procedure 45
      1.7.   Results... 45
          1.7.1.   Pick-and-place performance 46
          1.7.2.   Task completion time... 46
          1.7.3.   Excessive force application.. 47
          1.7.4.   Survey results... 48
      1.8.   Discussion 49

2. AN INVESTIGATION OF FORCE FEEDBACK DISPLAYS AND THEIR EFFECT ON TASK PERFORMANCE DURING A TELEOPERATED ABRASION TASK 51
   2.1. The effect of force feedback on force application 51
   2.2. The effect of force feedback on other performance aspects 52
   2.3. Force sensory substitution displays.. 53
      2.3.1. Theory on the effects of force feedback and force sensory substitution displays on task performance 54
      2.3.2. Empirical findings on the effects of force feedback and force sensory substitution displays on task performance 55
   2.4. Research aim 56
   2.5. Research hypotheses 57
   2.6. Method 58
      2.6.1. Participants.. 58
      2.6.2. Experimental apparatus... 58
      2.6.3. Experimental design.. 59
      2.6.4. Procedure 60
   2.7. Results.. 61
      2.7.1. Excessive force application... 61
      2.7.2. Regulation of applied forces 62
      2.7.3. Force application precision 63
      2.7.4. The relationship between task completion time and measures of force 64
      2.7.5. Survey results 65
   2.8. Discussion 65
3. GUIDELINES FOR THE IMPLEMENTATION AND EFFECTIVE USE OF HAPTIC FEEDBACK FROM THE REMOTE ENVIRONMENT... 68

CHAPTER III. THE EFFECT OF HAPTIC GUIDANCE ON TASK PERFORMANCE 69

1. AN INVESTIGATION INTO THE EFFECT OF HAPTIC EXPERT DEMONSTRATIONS ON THE TASK PERFORMANCE OF NOVICE USERS WITH A TELEOPERATED MICRO-ASSEMBLY SYSTEM 70
   1.1. Theoretical approaches to motor skill training 70
   1.2. Haptic feedback in training simulations.. 71
   1.3. Haptic expert demonstrations... 72
   1.4. Research aim 75
   1.5. Research hypotheses 76
   1.6. Method 76
      1.6.1. Participants.. 76
      1.6.2. Experimental apparatus... 77
      1.6.3. Experimental design.. 77
      1.6.4. Procedure 78
   1.7. Results.. 79
      1.7.1. Timing of movements 79
      1.7.2. Task performance improvement 81
      1.7.3. Implicit task knowledge 84
      1.7.4. Survey results 85
   1.8. Discussion 86
2. AN INVESTIGATION INTO THE EFFECT OF HAPTIC ASSISTANCE FUNCTIONS ON TASK PERFORMANCE AND USER PERCEPTION WITH A VIRTUAL TELEOPERATION SYSTEM 90
   2.1. Previous studies on the effects of haptic assistance on task performance 91
   2.2. Haptic assistance vs. user autonomy.. 92
   2.3. Research aim 94
   2.4. Research hypotheses 94
   2.5. Method 95

- 2.5.1. Participants 95
- 2.5.2. Experimental apparatus 95
- 2.5.3. The assistance functions 96
- 2.5.4. Experimental design 97
- 2.5.5. Procedure 98
- 2.6. Results 98
  - 2.6.1. Perceived control over TOP movements 99
  - 2.6.2. Objective task performance measures 100
  - 2.6.3. Comfort assessments 103
  - 2.6.4. Subjective task performance assessments 104
- 2.7. Discussion 104
- 3. GUIDELINES FOR THE IMPLEMENTATION AND EFFECTIVE USE OF HAPTIC GUIDANCE ... 107

CHAPTER IV. META-ANALYTIC INVESTIGATIONS INTO TASK FACILITATIVE EFFECTS OF HAPTIC SIGNALS 108

1. PREVIOUS EFFORTS TO SYNTHESISE RESEARCH ON HAPTIC INTERFACES 109
2. META-ANALYSIS AS AN INVESTIGATIVE TOOL 111
3. DATA SELECTION CRITERIA 112
    - 3.1. Research quality 112
    - 3.2. Construct definitions 113
      - 3.2.1. Haptic applications 114
      - 3.2.2. Task performance 114
    - 3.3. Type of effect 115
4. METHOD 116
    - 4.1. Literature search 116
    - 4.2. Inclusion criteria 116
    - 4.3. Effect size calculations 117
    - 4.4. Analyses 117
    - 4.5. Procedure 117
5. RESULTS 119
    - 5.1. Vibrotactile feedback 119
    - 5.2. Force feedback 120
    - 5.3. Haptic expert demonstrations 121
    - 5.4. Haptic assistance 121
6. DISCUSSION 123
7. GUIDELINES FOR THE IMPLEMENTATION AND EFFECTIVE USE OF HAPTIC SIGNALS 127

CHAPTER V. THE HMI DESIGN GUIDE 128

1. PREVIOUS COMPILATIONS OF GUIDELINES FOR THE DESIGN OF HUMAN-MACHINE INTERFACES 128
    - 1.1. Multicriteria Assessment of Usability for Virtual Environments (MAUVE) 130
    - 1.2. The Presence Design Guide 130
2. DEVELOPMENT OF THE HMI DESIGN GUIDE 132
    - 2.1. Collected guidelines 132
    - 2.2. Implementation 132
    - 2.3. Structure 132
      - 2.3.1. Work domain 134
      - 2.3.2. Prioritisation of work output requirements 135
      - 2.3.3. Task analysis 136
      - 2.3.4. Specification of contextual factors 137
      - 2.3.5. End-user specifications 138
      - 2.3.6. Output 139

2.4.	An Example: Interface design for teleoperated minimally-invasive surgery according to the HMI Design Guide.	140
3.	DISCUSSION	141

CHAPTER VI. SUMMARY AND GENERAL DISCUSSION 142

OUTLOOK 148

REFERENCES 150

APPENDICES 166

Figures

Figure 1. Illustration of a teleoperation set-up. 9

Figure 2. A schematic overview of processes involved during human-machine interaction. . 15

Figure 3. Schematic depiction of mechanoreceptors in the human hand. 18

Figure 4. The GROPE-III haptic display in use. 22

Figure 5. The TeslaTouch. 23

Figure 6. The ViSHaRD 10. 23

Figure 7. The Feelex I. 24

Figure 8. The SAM exoskeleton prototype. 24

Figure 9. Examples of serial, parallel, hybrid mechanisms. 25

Figure 10. An example of a parallel-kinematic design: the Quanser 5-DOF Haptic Wand. ... 25

Figure 11. Two examples of impedance controlled haptic interfaces. 26

Figure 12. The admittance-controlled HapticMaster. 27

Figure 13. The micro-assembly teleoperator setup. 42

Figure 14. Operator workplace. 43

Figure 15. Visual feedback from the remote environment. 43

Figure 16. Respective mean percentages of successful pick and place movements of all attempted trials. 46

Figure 17. Mean task completion times and their standard deviations for trials conducted with and without vibrotactile feedback. 47

Figure 18. Mean peak forces and their standard deviations for picking and placing acts with and without vibrotactile feedback. 48

Figure 19. Kuka KR100 Robot with ball bearing end-effector. 59

Figure 20. Depiction of the visual force substitution display.. 60

Figure 21. Average maximum forces for trials with haptic force feedback and trials without haptic force feedback. 62

Figure 22. Average force standard deviations for trials with force feedback and trials without force feedback. 63

Figure 23. Mean forces for trials with force feedback and trials without force feedback. 63

Figure 24. Mean task completion times and standard deviations for trials with force feedback and trials without force feedback in seconds. 65

Figure 25. Operator and teleoperator setup. 77

Figure 26. Track demonstrated by expert and novel track. 78

Figure 27. Left: demonstration by expert in the visuo-haptic demonstrations group. Right: teleoperator and test tracks. 79

Figure 28. Mean time deviations from the expert's demonstration. 80

Figure 29. Position inaccuracy in task performance. 82

Figure 30. Mean task completion times with standard deviations for the track previously demonstrated by the expert. 84

Figure 31. Mean task completion times with standard deviations for the novel track. 84

Figure 32. Number of people stating whether they believed the expert's demonstration helped them. 86

Figure 33. Operator setup with ViSHaRD7 and Head-Mounted Display 95

Figure 34. Screenshot of the virtual maze. 96

Figure 35. Mean values and standard deviations for estimates of control over TOP movement for each assistance function condition. 100

Figure 36. Mean times spent in contact with a wall and corresponding standard deviations for each assistance function and the control condition. 101

Figure 37. Mean task completion times and standard deviations for each trial and assistance function. 102

Figure 38. Screenshot of the Welcome Screen of the HMI Design Guide. 133

Figure 39. Screenshot example of the specification of work domain. 134

Figure 40. Screenshot example of the specification of performance goals. 135

Figure 41. Screenshot example of the task analysis. 136

Figure 42. Screenshot example pertaining to the specification of contextual requirements and system specifications. 137

Figure 43. Screenshot example showing the specification of end-user characteristics. 138

Figure 44. Screenshot example of the HMI Design Guide recommendations. 139

Tables

Table 1. Scale values for ratings of perceived comfort in working with each type of assistance. 103

Table 2. Scale values for ratings of perceived task performance quality. 104

Table 3. Summary of random-effects meta-analysis results for the comparison "vibrotactile feedback present" vs. "vibrotactile feedback absent". 119

Table 4. Summary of random-effects meta-analysis results for the comparison "force feedback present" vs. "force feedback absent". 120

Table 5. Summary of random-effects meta-analysis results for the comparison "haptic expert demonstrations" vs. "no haptic expert demonstrations". 121

Table 6. Summary of random-effects meta-analysis results for the comparison "haptic assistance present" vs. "haptic assistance absent". 122

Table 7. System design standards. 129

Table 8. Mean values and deviations of average force standard deviations (N) for each trial.167

Table 9. Mean values and standard deviations of mean force values (N) for each trial conducted with and without force feedback. 167

Table 10. Pearson correlation coefficients (r) and probability values (p) for force measures (N) task completion time (sec.) by feedback type. 168

Table 11. Mean values and deviations of task completion times (sec.) for each trial. 168

Table 12. Results of one-sample t-tests comparing mean negative deviations to the task completion time of the expert. 169

Table 13. Means, medians, and standard deviations for successful pick-and-place attempts (percentage)... 170

Table 14. T-tests and probability values with effect sizes for each comparison of control ratings between different types of assistance. 171

Table 15. T-tests and probability values with effect sizes for each comparison of the collision times between different types of assistance. 171

Table 16. T-tests and probability values with effect sizes for each comparison of the task completion times between different types of assistance. 172

Chapter I.

An Introduction to Teleoperation Systems and Haptic Interfaces

1. What is a teleoperation system?

Teleoperation systems are designed to combine human decision-making abilities with mechanical properties of strength and endurance. By extending a person's sensing and manipulation capabilities to a remote environment, teleoperation systems allow humans to perform tasks in an inaccessible or hazardous environment without putting themselves at risk. Furthermore, coupling humans with robotic systems enables humans to overcome some of their physical limitations as they may enable them to perform tasks at submicroscopic levels or provide them with superhuman strength.

1.1. General set-up of a teleoperation system

Although many variations exist, at its core, each teleoperation system is comprised of the same basic components. Central to the idea of teleoperation is the linking of two distinct spatial domains, which will from now on be referred to as the operator environment and the teleoperator environment (see Figure 1).

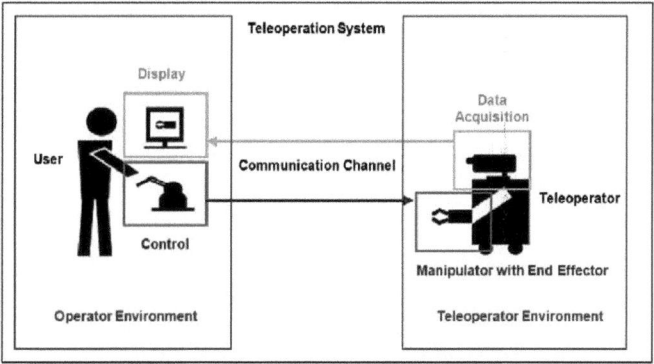

Figure 1. Illustration of a teleoperation set-up.
Source: adapted from Pongrac (2008), p. 7.

In the operator environment, a human user interacts with the controls and displays comprising the human-machine interface of the operator device. A communication channel links the operator to the teleoperator device in the remote environment, where it controls the teleoperator's manipulator via actuators. The manipulator is usually equipped with end-effectors, such as clamps or grippers that can interact physically with objects in the remote environment. In addition, information that is recorded by sensors in the remote environment can be transmitted to the operator side, where it is displayed to the human user.

1.2. Definition of terms

The terms master and slave devices are often used to denote operator and teleoperator devices, respectively. In the following work, the terms operator/master and teleoperator/slave will be used interchangeably. In some cases, a virtual simulation replaces the physical teleoperator, as these tend to be cheaper and are usually implemented more easily. It might be argued that virtual systems are not teleoperation systems in the strictest sense, as they do not allow a human to physically perform a task in a remote environment. Yet, virtual reality (VR) simulations are extensively relied upon in the development of teleoperation systems, as well as in the training of operators. The borders between virtual and real teleoperation systems are further blurred when one considers that many physical teleoperation systems incorporate augmented reality features and simulate sensory output displayed to the operator. Hence, for the remainder of this work, virtual systems that are linked in purpose and/or design to teleoperation systems are considered a particular type of teleoperation system.

Some researchers make a distinction between telerobotic systems, which use electronic components and purely mechanical telemanipulation systems. For this work, both types of systems are subsumed under the term "teleoperation system". Moreover, for the present work, the term teleoperation system will not distinguish between supervisory systems that are only used for observation and inspection of the remote environment (e.g. unmanned aerial vehicles), and manipulation systems that allow the user to directly affect the environment in some way. For the purpose of this work, a teleoperation system will therefore be defined broadly as:

> "a human/machine hybrid system, which continuously relies on human input for the control of a technological device (or its virtual representation) in a remote (including virtual) environment, and which is equipped with sensors and/or tools with the purpose of inspecting or manipulating the remote environment".

For example, according to this definition, a videoconference system would not classify as a teleoperation system, unless the movements of the cameras are remotely controlled by a human. Similarly, a remote-controlled toy car would not classify as a teleoperation system, unless it has sensors installed that provide information to its operator or unless it is purposefully used to physically affect the environment in some way.

2. Applications of teleoperation systems

Industrial teleoperation systems have been employed as early as the 1950s, when Ray C. Goertz began to develop his mechanical master-slave manipulation system at the Argonne National Laboratory with the aim of handling radioactive materials safely (Goertz & Thompson, 1954). In the 1960s, teleoperated manipulators had already been deployed in the decommissioning of a nuclear warhead of the US Army (Aracil, et al., 2007). In fact, to the day, the nuclear industry has remained one of the major driving sources behind the proliferation and development of industrial teleoperation systems.

Another main contributor to the development of teleoperation systems is the space industry. The Flight Telerobotic Servicer (FTS) program (McCain, Andary, Hewitt, & Haley, 1991) and the NASA/NBS Standard Reference Model (NASREM) (Albus, McCain, & Lurnia, 1989) are only two of the many contributions of the US National Space and Aeronautics Administration (NASA). A summary of NASA's telerobotics programs can be found in the 21st Century Guide to Robotics (NASA-JPL, 2004). Numerous contributions to research and development of telerobotic systems were also made by the German Aerospace Center (DLR) (Hirzinger, Brunner, Landzettel, Sporer, Butterfaß, & Schedl, 2003), the Canadian Space Agency (CSA) (Piedboeuf & Dupuis, 2001) and the Japanese Space Agency (NASDA) (Iwata T., 2001), amongst others.

Over the past decades, the employment of teleoperation systems has spread to numerous other domains. Teleoperated vehicles have been employed in warfare (Laird, Bruch, West, Ciccimaro, & Everett, 2000) and underwater exploration (Utsumi, Hirabayashi, & Yoshie, 2002). Robot-assisted teleoperation systems are frequently applied to surgery (Rosen, Hannaford, & Satava, 2010), and attempts have been made to use teleoperated systems for live-line maintenance (Aracil, et al., 1995), search & rescue operations (Jacoff, Messina, Weiss, Tadokoro, & Nakagawa, 2003), education (Yang, Chen, Petriu, & Petriu, 2004), and care of the disabled and the elderly (Kawamura & Iskarous, 1994). Due to the enormous variety in employment purposes of modern teleoperation systems, in-depth evaluations will necessarily have to focus on a small subset of systems. Focus of the present work are teleoperation systems that are employed for strictly industrial purposes, which include maintenance, transportation and assembly work.

2.1. Benefits of the industrial employment of teleoperation systems

Undeniably, teleoperation systems offer distinct advantages over unmediated, hands-on approaches for a wide range of industrial applications. For instance, in minimally invasive surgery, teleoperation systems can increase ergonomic comfort for the surgeon whilst reducing human-induced risks such as surgeon tremor (Stylopoulos & Rattner, 2003). Offering a workplace which is ergonomically well-designed and increases the operator's (theoretical) ability to perform high precision movements has also been found beneficial for the assembly of micro-parts (Zaeh & Reiter, 2006). Furthermore, operating the system from a distant environment reduces the risk of contamination, thus lowering fatality rates for surgery and assuring high production quality in micro-assembly (Zaeh & Reiter, 2006). The nuclear industry is among the main contributors to the field of teleoperation system development, as these systems enable humans to handle radioactive material without the risk of health-damaging and potentially lethal consequences of prolonged or intense exposure (Aracil, et al., 2007). Space agencies are interested in the development of sophisticated teleoperation systems that would allow their users to perform maintenance on satellites while they are still in orbit (on-orbit servicing), thus reducing the need for dangerous and expensive astronaut servicing (Reintsema, Landzettel, & Hirzinger, 2007).

It is still debated whether teleoperation systems constitute viable alternatives to automated systems. By their nature, teleoperation systems are necessarily less accurate and more error prone than automated robotic systems as they rely on the correct interpretation and reaction of the human user. However, when tasks are rarely executed or require flexibility, teleoperation systems may be the only viable option for task completion, as automation may not be practically or financially feasible under these circumstances (Book & Love, 1999).

2.2. Current challenges in the industrial employment of teleoperation systems

Despite the many advantages of teleoperation systems, working with currently available systems poses a number of challenges for large-scale industrial deployment. For one, it is much more difficult to coordinate the movements of a teleoperated device than it is to coordinate one's limb movements. In fact, Zhai & Senders (1997) observed that around one quarter of all movements with a multiple-degree of freedom (DOF) system is uncoordinated. Deml (2004) found in a more challenging task that only two-thirds of movements can be considered coordinated. As a result, it takes generally much more time to perform a task with a teleoperation system than it would if the task was performed manually (Geiger, Popp, Färber, Artigas, & Kremer (2010); Adams, Klowden, & Hannaford (2001); Unger, Berkelman, Thompson, Lederman, Klatzky, & Hollis (2002)). Another consequence of the poor coordination of the teleoperator's movements is that performance tends to be error prone (Ware & Balakrishnan (1994); Massimino, Sheridan, & Roseborough (1989)). A second challenge remains the precise regulation of forces that are applied in the event of contact with an object or surface in the remote environment. Thus, there is a distinct risk that the teleoperator itself and any material in the remote environment that comes into contact with the teleoperator, might be damaged due to excessive forces applied by the user (Radi, Reiter, Zaidan, Nitsch, Faerber, & Reinhart (2010); Tavakoli, Patel, & Moallem (2005)). Safety mechanisms that automatically incapacitate the teleoperator in the event of excessive forces can be implemented into most teleoperation systems; however, frequent triggering of these mechanisms would further delay work performance and may not always prevent damage to the material.

The causes of these performance impairments are only partially understood. While the problem of poor movement coordination has been observed in numerous cases, only few attempts have been made to discover the causes of this impairment. Deml (2004) suggested that it might be linked with the complex kinematic of particular devices, as humans experience difficulty in the coordination of rotary movements. This concurs with Aracil, et al. (2007), who argued that kinematic transformations might lead to a perceived discordance between operator and teleoperator movements. The resulting incoherence between the user's commands and their execution may disrupt the user's ability to judge the relative position of the teleoperator arm to other objects as well as their ability to anticipate future positions of the manipulator. Furthermore, a likely cause of users' general inability to apply teleoperator forces accurately and appropriately may be the lack of tactile feedback from the remote environment, which informs the user of material properties such as compliance, and which, in the absence of unambiguous visual cues, is essential to human force regulation (Birbaumer & Schmidt, 2003).

Certainly, other aspects of work performance suffer in teleoperated applications, aside from those just mentioned. For instance, during telesurgery, a critical problem is the inability to differentiate between tissues of different texture and compliance, as a result of the lack of tactile information during remote manipulation. A large body of research already exists on the (in)ability of users to make correct qualitative judgements based on perceived object attributes in teleoperation systems (for summaries, see e.g. Tan, Eberman, Srinivasan, & Cheng (1994); Klatzky & Lederman (1999); Jones & Sarter (2008)). Hence, this issue will not be further considered in this work. Instead, the focus of this work will be on quantitative aspects of work performance, specifically on movement coordination, as indicated by timing and movement errors, as well as on measures of human force application.

In principle, there are two ways of increasing the effectiveness of any human-machine interaction: training of the user and enhanced system design. Although it has been demonstrated that work performance with teleoperation systems may improve with increased training of the user (e.g. Adams, Klowden, Hannaford (2001); Chen, (2008)), this tends to be a laborious and subsequently costly process - oftentimes with limited success. Hence, it seems vitally important to explore ways of improving work performance with teleoperation systems by improving the design of system components that are the most likely cause of the observed performance problems.

Considering that robots and humans are both quicker and more accurate in their actions when working independently, it seems likely that the root of this inefficient collaboration between human and machine can be traced to the design of the human-machine interface. Consequently, improved design of this system component is likely to lead to superior user performance with teleoperation systems. To this end, research is needed that investigates in what way work performance with teleoperation systems may be affected by interface design, so that guidelines can be established which may advise system developers on the design of effective interfaces. The following section will outline the function of the human-machine interface in teleoperation systems, present common designs and discuss current approaches to interface research and development, before research is presented that aims to address this need.

3. The human-machine interface

As teleoperation systems aim to capitalise on the synthesis of human flexibility with robotic physical capabilities, the human-machine interface is arguably the most important component of a teleoperation system. At the same time, constructional and functional requirements of this component are difficult to define. In effect, the human-machine interface is responsible for translating between human and machine so that their respective actions can be coordinated. In order to guarantee an effective translation process, the following main processes need to be taken into account during human-machine interaction (see Figure 2):

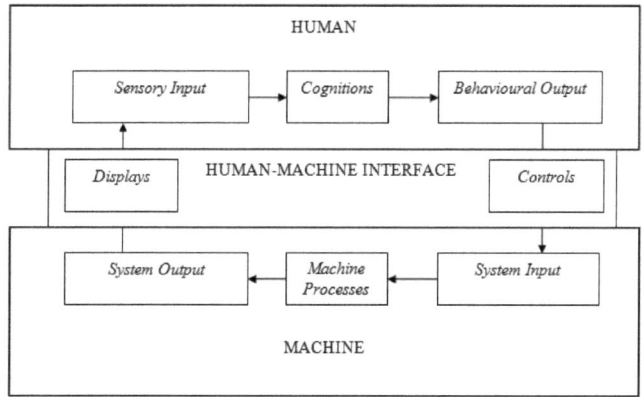

Figure 2. A schematic overview of processes involved during human-machine interaction.
Source: adapted from Glendon, Clarke, McKenna (2006), p. 136.

On the most basic level, a machine interacts with a human by displaying visual, auditory, tactile or other types of information to the user. The user then attends to those sensory inputs considered relevant. The attended information is then organised and interpreted based on different cognitions such as memories and decision making abilities. When the decision is made to perform a particular action, this decision is translated into behaviour, which is constrained by available effectors, like muscles, and kinetic ability. The behavioural output then acts on the machine controls. The controls have to transform this input into data that can be interpreted by the machine, so that it can execute the user's order, and produce output for the operator, from which point the cycle continues. Naturally, this model is very simplified and represents only a crude approximation to the processes involved in a human-interaction. It should also be noted, that the direction of the information flow depicted in this figure is rather deceptive, as it does not necessarily flow uni-directionally. Many instances have been recorded in which our memories and expectancies inform our sensory perception, before they affect behavioural output. Although this concept of human-machine interaction is imprecise, it illustrates an important point: the effectiveness of the human-machine interaction is determined by the degree to which the translation of human behavioural output into system input AND the conversion of system output to human sensory input is successful. However, equally important are cognitive processes which determine the most appropriate course of action based on the provided input.

Simply put, a human-machine interface is well-designed, if it manages to keep the cognitive effort needed to interpret information, decide upon the correct course of action and transform this action into appropriate machine input to an absolute minimum (Hedicke, 2000). This is only possible if the affordances of the control are suitable to the purpose of the action, and if the displays inform of the current state whilst allowing for an accurate prediction of future states. The more cognitive effort is required, the less effective the interaction and the greater the potential of operation mistakes.

3.1. Multi-modal interfaces

Multi-modal interfaces, which incorporate more than one sensory modality into controls and displays, are widely believed to enhance the effectiveness of the translation process that takes place during human-machine interaction. Different views are taken on why that may be the case. Some have ventured that the redundancy of in-/output channels renders the translation process less vulnerable to "misinterpretation" (Lewkowicz & Kraebel, 2004), as irregularities in the incoming stimulation or in the subsequent processing generally only affect one sensory modality (Bertelson & De Gelder, 2004). As such, multi-modal sensory input cannot only resolve inherent ambiguities that exist within one particular sensory modality but also those resulting from conflicting information of two or more sensory modalities (Allison & Jeka, 2004).

Others have suggested that interaction with a remote environment via multi-modal in- and output options more closely approximates a non-mediated interaction with the environment than e.g. unimodal displays, thereby rendering the human-machine interaction more intuitive (Chu, Dani, & Gadh, 1997). In fact, a natural predisposition of human senses towards certain types of information has been documented (Muthig, 1990). For instance, the visual system was found to be particularly adept at interpreting colour, size, shape, movement and distances between objects. The auditory system is oriented towards pitch, volume and direction and movement of sound signals. The haptic sensory system is specialised with regard to the successive and simultaneous comprehension of weight, texture, compliance, temperature, pressure, and movement and relative position of limbs (Muthig, 1990).

The majority of modern multi-modal interfaces speak predominantly to the visual and auditory senses. Increasingly, solutions to the performance problems with teleoperation systems are advanced that focus on the incorporation of the haptic sensory modality into the human-machine interface.

3.2. Haptic human-machine interfaces

3.2.1. What exactly is "haptics"?

In 1892, Max Dessoire introduced the term "haptic" as the "science of the human touch" (Jütte, 2008), derived from the Greek word *απτός*, meaning tactile or tangible. Definitions given for the term haptic vary widely in the scientific community. Gibson (1966) defined the haptic sense as:

> "The sensibility of the individual to the world adjacent to his body by use of his body" (Gibson, 1966, p. 97).

Klatzky & Lederman (2003) distinguish cutaneous, kinaesthetic and haptic perception as subcategories of the sense of touch. In reference to the terminology used by Loomis & Lederman (1986), they further elaborate:

> "[...] the cutaneous system receives sensory inputs from mechanoreceptors- specialised nerve endings that respond to mechanical stimulation (force) that are embedded in the skin. The kinaesthetic system receives sensory inputs from mechanoreceptors located within the body's muscles, tendons, and joints. The haptic system uses combined inputs from both the cutaneous and kinaesthetic systems [and] is associated in particular with active touch." (Klatzky & Lederman, 2003, p. 148).

Aiming to maintain some form of continuity with previous work on haptic interfaces, for the present work, haptic perception will be defined broadly as the "sense of touch and limb position and movement", whereby further distinctions are made between cutaneous/tactile perception, defined as the sense of touch, and kinaesthetic/proprioceptive perception, defined as the sense of limb movement and of the limbs' relative position in space.

3.2.2. The physiological basis of haptic perception

A haptic percept is the result of integrated sensory information and is formed based on the sensitivity and conductivity of various sensory cells embedded in skin, subcutaneous tissue, joints and muscles. The receptors of the skin can be classified into four distinct groups, based on their response to stimulation and the spatial characteristics of their receptive fields (Goodwin & Wheat, 2008). Slowly-adapting (SA) receptors respond as long as a stimulus is in place, whereas fast-adapting (FA) receptors respond only briefly to the beginning and end of stimulation. The receptive fields of Type I afferents are small and feature multiple zones of maximum sensitivity and rapidly increasing thresholds towards its borders. In contrast, Type II afferents feature larger receptive fields with less well-defined borders, as the thresholds increase only gradually towards the edges (Goodwin & Wheat, 2008).

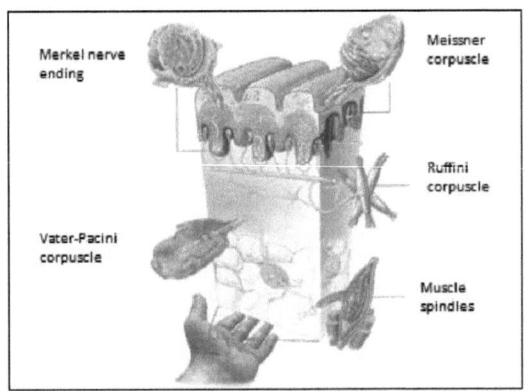

Figure 3. Schematic depiction of mechanoreceptors in the human hand.
Source: Halata & Baumann (2008), p. 92.

Merkel cells are slowly-adapting Type I (SAI) cutaneous mechanoreceptors, specialising in the monitoring of pressure exerted on the skin surface (Halata & Baumann, 2008). They are also involved in the perception of form and texture (Hendry, Hsiao, & Bushnell, 1999). Like the Merkel cells, Meissner cells (FAI) are also found in the superficial cutaneous tissue (see Figure 3). These perceive movement along the skin, and are particularly involved in the grasping of objects, as they allow for the adjustment of grip forces until the object no longer slips over the skin (Hendry, Hsiao, & Bushnell, 1999). The Ruffini corpuscles (SAII) detect stretching of the dermis or ligaments and joint capsules, whereas the Pacinian corpuscles (FAII) sense vibrations (Halata & Baumann, 2008). Both types of receptors are settled in deeper tissue of the skin. Kinaesthetic receptors like muscle spindles and Golgi tendon organs provide information required for the control of length and tension of muscles.

In addition to the information provided by the various mechanoreceptors previously described, temperature changes and painful stimuli are mainly detected through free nerve endings (polymodal nociceptors) distributed throughout the skin and connective tissue of the locomotion apparatus (Halata & Baumann, 2008). The perceptive and output capabilities of the human hand and arm have been studied extensively in the past. Detailed specifications can be found in Kern (2009, pp. 51-54).

Although haptic perception also encompasses the sensing of temperature and pain, the present work will focus on the perception of movement and forces, as it stands to reason that these percepts are most relevant to the previously outlined problems of poor movement efficiency and excessive force application in the use of industrial teleoperation systems.

3.3. The role of haptic feedback in human-machine interaction

Over the past two millennia, the haptic sense has received mixed accolade. Whilst early philosophers like Aristotle (384 - 322 B.C.) thought the sense of touch to be the least important sense, later scholars such as Thomas Aquinas (1224/5 - 1274 A.D.) believed touch to be the most dominant and invariably the most important of all senses - a view which can also be encountered in the works of other Christian, Muslim and Jewish scholars of the Middle Ages (Jütte, 2008). Today, most researchers would agree that, while the visual sense tends to dominate human perception under most circumstances, haptic perception is nonetheless pivotal to efficient human functioning (Bertelson & De Gelder, 2004). This suggests that it might also be an important factor in effective human-machine interaction. However, while the literature is replete with empirical investigations into the effects of various haptic displays on human haptic perception (e.g. Klatzky, Lederman, & Metzger (1985); Tan, Eberman, Srinivasan, & Cheng (1994); Jones & Sarter (2008)), potential causal relationships between haptic feedback and effective human-machine interaction have been largely inferred from observation rather than theoretical deduction.

Yet, the literature on human perception can provide some clues as to why a lack of haptic feedback might impair human performance with a teleoperation system. Although it was estimated that humans tend to perceive more than 90% of their environment through their eyes (Mauter & Katzki, 2003), certain object properties, including weight, compliance, viscosity, and texture are also perceived through the haptic sense (Klatzky, Lederman, & Reed (1987); Klatzky & Lederman (2003)). Moreover, humans tend to rely on tactile information when they manipulate an object, for example, when grasping an object or when placing it onto a surface. For instance, the human adaptation of grip forces to external requirements is caused by polysynaptic reflexes, which allow humans to anticipate intuitively the amount of pressure they need to apply in order to achieve their task goal (Birbaumer & Schmidt, 2003). In fact, clinical studies have shown that patients with total loss of their tactile and proprioceptive senses are unable to drink out of soft plastic cups, since they are unable to grasp and hold on to the cup (Birbaumer & Schmidt, 2003). Similarly, when placing an object onto a surface, we combine information from our memory regarding the weight, compliance of the object and the friction of the surface (Ballesteros, 2008), with the information we receive from our tactile sense (e.g. through cutaneous receptors detecting skin deformation) and our visual sense (e.g. in judging the distance between the object and the surface, or when observing the indentation of object or surface) (Lederman & Klatzky, 2009). In fact, Ernst & Banks (2002) proposed a statistical model, which assumes that the more precise modality will receive a higher weighting than the less precise modality as the human tends to minimise the variance of the final estimate. Empirical evidence seems to support this notion (Gentaz & Hatwell, 2008). This would suggest that haptic perception becomes more important, the less reliable visual information is available.

During the remote handling of objects, haptic information about the remote environment is largely unavailable and needs to be compensated through the reliance on accumulated experience and other sensory modalities, typically vision. If the loss of information due to the unavailability of the haptic sense cannot be fully compensated, task performance would likely suffer as a result. This is particularly evident in teleoperated surgery, for which excessive force application, long contact times with tissue, and difficulty in the differentiation between materials of differing softness are widely reported problems (Perreault & Cao, 2006). Similarly, the overreliance on the visual modality introduces a risk of error in the interpretation of the visual signals, particularly when important cues of depth are missing. The consequences of possible signal misinterpretation can, for instance, be observed during teleoperated assembly, where placement errors in pick-and-place tasks can be problematic (Debus, Jang, Dupont, & Howe, 2002).

The theoretical considerations laid out thus far imply that haptic interfaces can ameliorate the previously observed performance problems with teleoperation systems. Displays that simulate tactile information otherwise lacking may adequately compensate for this loss of information from the remote environment and thereby enhance the user's ability to control applied forces precisely, thus reducing the risk of excessive forces on the teleoperator side. By providing cues of depth in a two-dimensional depiction of the remote environment (e.g. via video feed), simulated haptic feedback may also improve the user's ability to coordinate the teleoperator's movements by resolving sensory ambiguity. However, as a bidirectional system, the haptic modality can be used to affect the user's movements much more directly as haptic input/output devices can be used to guide the user's movements actively. Several approaches have been conceived of in order to improve the user's movement coordination ability. For example, haptic guidance systems have been used to prevent users from entering particular areas (Joly & Andriot, 1995), pull them along pre-defined paths (Pezzementi, Okamura, & Hager, 2007), or simply to demonstrate effective movements haptically (O'Malley, Gupta, Gen, & Li, 2006).

3.4. Haptic human-machine interfaces: an overview

In a sense, most human-machine interfaces already incorporate tactile and kinaesthetic features. In some form or other, a user's touch and movement almost always provides the input for conventional systems, be it in the form of touch screens, through the push of a button or pulling of a lever. Yet, with conventional interfaces, system output is typically delivered in the form of visual and auditory signals. Unlike vision and audition, however, the haptic system is bidirectional as humans send and receive haptic signals (Tan H. Z., 2000). Certainly, many gadgets already tap into this use of the tactile perception channel as they send out vibratory signals in order to attract immediate attention. For instance, "silent" alarms of mobile phones and rumble packs in gaming controls have long been accepted and integrated into our daily routines. More advanced haptic interfaces make use of force feedback devices that cannot only measure the position and contact forces of the user's hand, but also feedback position and/or force signals to the user (Tan, Eberman, Srinivasan, & Cheng, 1994). Hence, haptic interfaces allow for human-machine interaction through touch and mostly in response to user movements.

3.4.1. The origins of haptic human-machine interfaces for teleoperation systems

The telemanipulation system developed by Goertz & Thompson (1954) at the Argonne National Laboratory is widely cited as the first robotic teleoperation system to use force feedback. As early as 1965, Ivan Sutherland, a pioneer of computer graphics, proposed an "ultimate display for virtual applications" that included haptic feedback with the aim of enhancing human-machine interaction (Sutherland, 1965). It is said that this vision marked the origin of project GROPE, one of the first systematic research projects on haptic interfaces for teleoperation/virtual applications, which was conducted by Frederick Brooks, Jr. and his associates at the University of North Carolina at Chapel Hill in 1967 (Burdea, 1996). Its aim was to develop a visual-haptic display for the real-time simulation of three-dimensional molecular docking forces, which was used to teach students in an introductory physics course. Its final system, GROPE-III which was presented more than two decades later (see Figure 4), demonstrated that force feedback facilitated the task of assigning the lowest potential energy to docked protein molecule combinations as it furthered the understanding of underlying principles of physics (Brooks, Ouh-Young, & Batter, 1990).

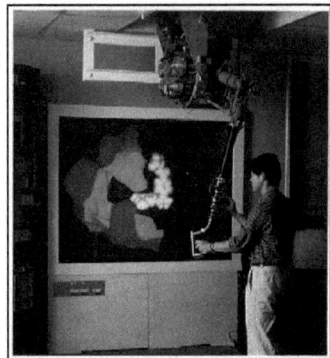

Figure 4. The GROPE-III haptic display in use.
Source: Brooks, et al. (1990), p. 177.

3.4.2. Classifications of haptic interfaces

Since its beginnings, decades of research and experimentation on the construction and application of haptic interfaces have produced an enormous variety of haptic interfaces that are available today, many of which show a potential of improving the human-machine interaction in teleoperation systems. Overviews of different haptic systems have been provided by a number of authors (e.g. Iwata (2008); Burdea (1996); Kern (2009); Martin & Savall (2005); Laycock & Day (2003)). In an attempt to synthesise current research efforts, a number of different classification schemes have been proposed, almost all of which tend to emphasise particular technical properties of individual devices.

For instance, a broad distinction between different interfaces is often made based on their predominant feedback component. Haptic interfaces with a predominantly tactile feedback component stimulate skin sensations, which are primarily used to convey information regarding specific surface properties of objects such as texture, temperature and friction (Iwata H. , 2008). For example, in a cooperation of Disney Research and the Carnegie Mellon University in Pittsburgh, Bau, Poupyrev, Israr & Harrison (2010) developed the TeslaTouch, which uses electrovibration technology to enhance touch interfaces with tactile feedback, for instance regarding a virtual object's texture (see Figure 5).

Figure 5. The TeslaTouch.
Source: Bau et al. (2010), p. 290.

In contrast, kinaesthetic devices are aimed at stimulating proprioception of the joints and muscles. An example of such a system is the ViSHaRD10 (Virtual Scenario **Haptic Rendering Device** with **10** actuated degrees of freedom). The hyper-redundant kinematic design of this interface aims at avoiding interior singularities, thus increasing the workspace while reducing the overall size of the device (see Figure 6). Furthermore, the actuated kinematic redundancies allow for a variable force output of up to 170 N (Ueberle, Mock, & Buss, 2007). Due to its configurative versatility, this device has the potential of serving as a benchmarking testbed that may aid the cost-effective development of haptic applications. The idea is that the workspace and/or force capability may be constrained to desired specifications to test the general applicability of a particular haptic application before a specialised display is developed (Ueberle, Mock, & Buss, 2007). To date, the ViSHaRD10 has mostly been used for research purposes.

Figure 6. The ViSHaRD 10 at the TU München.
Source: Buss, Kuschel, Lee, et al. (2006), p. 3.

Oftentimes, a distinction is further made between stationary systems, which are placed in fixed positions and are mechanically grounded to a desk or a wall, and portable systems, which are affixed to the human body. While stationary systems have the advantage of off-loading the actuator weight from the user, they necessarily restrict the user in their freedom of motion (Burdea, 1996). The ViSHaRD10 is an example of a kinaesthetic, non-portable system, which is mechanically grounded to the ceiling. The FEELEX (derived from the words feel and flex), represents a non-portable tactile system which was developed at the University of Tsukuba, Japan (see Figure 7). The inventors suggested a number of possible applications for this system ranging from palpation training of surgeons to interactive art (Iwata, Yano, Nakaizumi, & Kawamura, 2001).

Figure 7. The Feelex I (developed in 1997).
Source: Iwata, Yano, Nakaizumi, & Kawamura (2001), p. 471.

In contrast, portable interfaces are theoretically spatially less restrictive, but most of them are carried by the user and are therefore limited in their overall weight and volume (Burdea, 1996). A large category of portable systems constitute the exoskeletons, such as the SAM (Sensoric Arm Master) depicted in Figure 8, which is the result of a joint development project of the European Space Agency (ESA) and the Université Libre de Bruxelles (Letier, et al., 2008).

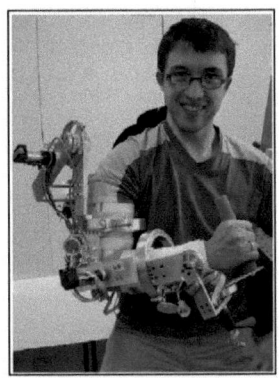

Figure 8. The SAM exoskeleton prototype.
Source: Letier, et al. (2008), p. 3504.

A further distinction can be made based on the underlying kinematics of multiple-DOF devices. In this context, kinematics refers to the mathematical calculation of movement patterns in mechanical systems (Kern, 2009). The kinematic structure of most haptic interfaces can be assigned to one of three main categories. In serial designs, the tool centre point (TCP) on the end-effector is connected to the base with actuated joints in a single kinematical chain; there are no passive joints (see Figure 9). The PHANTOM Omni by SensAble Technologies (Figure 11) would be one example of a serial-kinematic design.

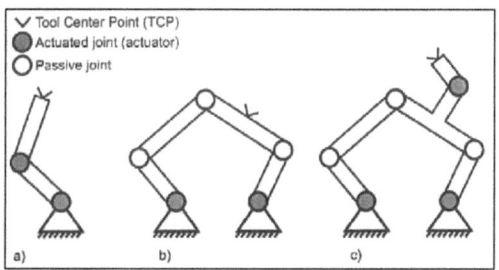

Figure 9. Examples of serial (a), parallel (b), hybrid (c) mechanisms.
Source: Kern (2009), p. 167.

In parallel designs, this link is established via a number of parallel kinematical chains. By placing all actuators at the frame (a rigid body), the moving masses can be minimised. Due to the small inertia, these types of design allow for quick motions (Tsumaki, Naruse, Nenchev, & Uchiyama, 1998). On the other hand, most serial mechanisms have a higher workspace-to-size ratio than parallel-kinematic structures (Kern, 2009). The 5-DOF Haptic Wand by Quanser (Figure 10) features a parallel-kinematic design.

Figure 10. An example of a parallel-kinematic design: the Quanser 5-DOF Haptic Wand.
Source: Quanser Industrials.

A special case is the hybrid kinematical design, which features both kinematic elements as it contains serial- and parallel-chain modules that are connected either in serial or in parallel. This type of design is often applied to haptic interfaces that are typically employed in multi-purpose rather than specialised teleoperation systems (Ueberle, 2006).

An alternative classification of kinaesthetic interfaces focuses on their internal structure, such as the featured control mechanisms. System designs of haptic interfaces may be either admittance or impedance controlled and closed-looped or open-looped controlled. Generally speaking, impedance-controlled systems produce a force as output, while the input is provided in form of position measurements. On the other hand, admittance-controlled systems generate output in the form of position changes, while the user provides input in form of a force reaction (Kern, 2009). An open-loop controller computes the input into the system using only the current state and its model of the system, but providing no feedback. In contrast, in closed-loop systems the output of the system is fed back to the input of the controller, thus providing a more accurate and adaptive, but generally also less stable form of control (Kern, 2009).

Partly due to their simple design, open-loop impedance controlled systems, such as the PHANTOM Omni by SensAble Technologies (see Figure 11), are amongst the most prolific devices available on the private market (Kern, 2009). In contrast, closed-loop impedance controlled systems, such as the delta.3 by ForceDimension are typically employed in research.

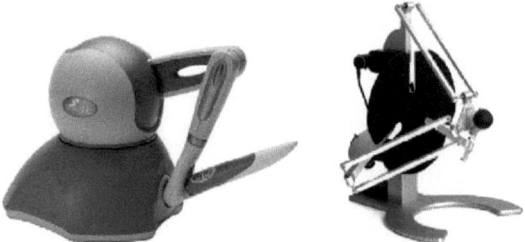

Figure 11. Two examples of impedance controlled haptic interfaces.
The open-loop controlled PHANTOM Omni (left) and the closed-loop controlled delta.3 (right).
Source: SensAble Technologies (left) and ForceDimension (right).

Open-loop admittance controlled systems are typically found in pin array-based tactile displays, such as the FEELEX shown in Figure 7. Closed-loop admittance controlled devices, such as the HapticMaster by MOOG (see Figure 12), are particularly advantageous for applications that require high stiffness, but they are invariably comprised of expensive components and the implementation of multiple DOF with these system structures can be technically quite challenging (Kern, 2009).

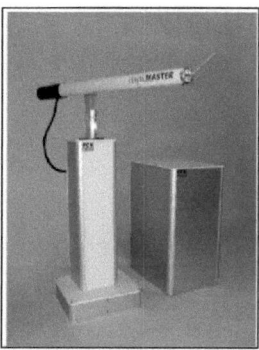

Figure 12. The admittance-controlled HapticMaster.
Source: Van der Linde, et al. (2002), p.1.

3.5. Current approaches to haptic interface research and development

Considering the wide variety of haptic technology at our disposal, the central questions driving this work are:

1. Which applications of haptic signals, if any, hold the potential of improving work performance with a teleoperation system?

2. In what way might they improve performance?

3. How can haptic interfaces be improved so that they optimise work performance with teleoperation systems?

For the investigation of these questions, a suitable framework, which guides the research and development of haptic human-machine interfaces needs to be adopted.

3.5.1. System-centred research and development

While the different haptic interfaces vary widely in their kinematic design, internal structure and the type of feedback that they provide, they all aim at communicating information that speaks directly to the human haptic sense. Yet, in the research and development of haptic interfaces, a distinctly system-centred approach currently prevails, whereby developers tend to focus their efforts on the improvement and development of specific mechanical properties in iterative stages of trial-and-error. Hereby, inventions and improvements are driven by technological affordances and feasibility rather than user or task requirements. The system-centred approach certainly has its merits, in particular during the early proto-type stages of development, where it is often considered unfeasible and impractical to conduct detailed task analyses and frequent user tests (Mao, Vredenburg, Smith, & Carey, 2005). Continuing this line of research and development from a system-centred perspective, however, means to risk that work performance with teleoperation systems will not be optimised, as it is exceedingly difficult with this approach to establish which, if any, components of an interface need to be augmented, and in what way these need to be further developed. This difficulty is attributable to a number of factors inherent to the system-centred approach to design.

For one, system-centred design fails to take into account inter- and intrapersonal variance in human performance with a particular system, which increases the risk that performance with a system that is observed during development, will not be comparable to that observed in the field. For instance, under this approach, the different iterations of a technological system are tested with a few selected individuals, typically the developer(s) themselves, all of which have detailed technical knowledge of the system. In comparison to elaborate tests with a larger sample of naive users, this testing process is time- and cost-efficient, and effective insofar as experts can provide specific design recommendations as well as suggest ways of implementing them. However, invariably, the responses of this highly selective, homogenous group of people are not necessarily representative of those obtained by a wider population (Bortz & Döring, 2006).

Secondly, whilst it would be unfair to claim that the role of the human in teleoperation systems has been ignored in system-centred design, previous efforts to incorporate the human into the design of the interface have typically focused on the physical characteristics of human and machine in-/output capabilities. The role of cognitions, which are responsible for perceiving and interpreting haptic signals and for determining the decision to perform particular behaviours, have been largely neglected in this context. These, however, are determined not only by perceived signals (bottom-up processing), but also by expectations (top-down processing) (Eysenck & Keane, 2000). Hence, by focusing on signal characteristics and human perception thresholds, we only know what humans *can* do but not necessarily what they *would* do in response to certain signal output. Human perceptions of and reactions to signals that are presented out of context, as it is frequently practiced during system-centred design, are therefore only marginally applicable to real-world applications.

Thirdly, in system-centred design, systems tend to be developed first, while the application of this technology and the purpose that it might serve is sought afterwards. On occasion, it might be appropriate and desired to adopt this approach in order to develop generalised systems that may later be employed in a number of different settings. However, in practice this means that during the development of these systems, these are often tested in a context that is very abstract and tailored to suit the technological capabilities and strengths of a particular system. If haptic interfaces are not tested under realistic conditions, it is difficult to establish how effective these interfaces might be in enhancing human-machine interaction in an industrial context and in what way they might be improved. As a result of their untested effectiveness in situations that reflect realistic task demands, many applications of haptic signals that show a potential for improving human-machine interaction with these systems never find their way to industrial large-scale deployment.

Finally, one might argue that system-centred approaches are ill-suited to provide a framework for the investigation of haptic human-machine interaction and its broad influence on task performance. This becomes evident from the extremely fragmented field of research, in which each study focuses on one particular system only. Comparative human performance studies observing task performance with several different devices are virtually non-existent. Since systematic empirical research, which investigates the impact of different haptic interfaces on human task performance with teleoperation systems, is sorely lacking, it has become particularly difficult to advise on the best use of haptic signals in teleoperation systems.

3.5.2. User-centred research and development

In contrast to the traditional system-centred approach to research and development, which is driven by considerations of technological feasibility, stands the user-centred approach, which is mainly concerned with improving the usability of new products (Mao, Vredenburg, Smith, & Carey, 2005). According to ISO 9241-11 usability is defined as "the extent to which a product can be used by specified users to achieve specified goals with effectiveness, efficiency and satisfaction in a specified context of use", although definitions vary widely in this field (Jokela, Iivari, Matero, & Karukka, 2003).

Similarly to the system-centred approach, user-centred design necessitates iterative stages of system design. During this iterative process, a developed product is assessed in a specified use context, whereby the results of this evaluation inform further product development. The improved product is then tested in further evaluations and continuously improved, until it is found to satisfactorily address pre-specified task and user requirements. Unlike system-centred design, user-centred design emphasises multi-disciplinary efforts of the technical and social professions, as well as the active involvement of users (Jokela, Iivari, Matero, & Karukka, 2003). As such, it takes intra- and interindividual differences in user behaviour into account, and ensures that the developed product meets task requirements that are likely to be encountered during the use of this product in real-life. In particular in the information systems industry, the user-centred design approach has increasingly gained support, as numerous studies have confirmed tangible benefits over system-center designed products (Mao, Vredenburg, Smith, & Carey, 2005).

In the development of emergent technologies, including haptic interfaces, user-centred design is rarely practiced, as many developers regard the design process as intimidating in its complexity (Nielsen, 1993), and some of its proposed design methods, such as focus groups and participatory design, were found to be ineffective and/or impractical in some circumstances (Mao, Vredenburg, Smith, & Carey (2005); Kujala (2003)). Moreover, since the bulk of previous work on the development of haptic interfaces has been system-centred, it would seem ill-advised to ignore these efforts. Instead, a new approach to interface design is sought that can build a bridge between system-centred and user-centred design.

3.5.3. Application-centred research and development

In this work, an application-centred approach to the research and development of haptic human-machine interfaces in industrial teleoperation systems, and consequently to finding answers to the central research questions, is proposed. Application-centred research and development uses existing technology and establishes its effectiveness within a specific use-context with users that are representative of the end user. Central to this approach are empirical investigations with representative samples of naive users and settings that reflect realistic task demands that aim at ascertaining which, if any, haptic applications might improve task performance, and under which circumstances these are effective.

Analysis of the haptic technology literature identified two main areas of application for haptic interfaces in teleoperation systems that have the potential of minimising current difficulties in the operation of telemanipulation systems:

(1) the simulation of haptic feedback from the remote environment and
(2) the application of haptic guidance signals to the operator control.

The focus of the present work is on these applications of haptic signals in industrial teleoperation systems and their potential of improving quantitatively measurable aspects of work performance that were found to be problematic in teleoperation, specifically human force regulation in the remote environment, and movement coordination as evident in task completion times and performance errors. Since it has been established that human cognitive processes need to be taken into account in the design of human-machine interfaces, user perceptions of usefulness and acceptance of the different applications are also of interest, as these cognitions may offer insight into performance behaviour with these systems, as well as provide indications for areas of further improvement.

A difficulty with the study of human cognitions during human-machine interaction is that these are not easily defined, observed or measured. In contrast, behavioural responses to signals allow for a distinctly quantifiable measurement of human-machine interaction. Behavioural measures also have the advantage that future behaviour can be fairly accurately predicted based on past behavior (Hogg & Vaughan, 2002). The derivation of design guidelines for interfaces that allow for effective work performance with teleoperation systems should therefore focus on behavioural measures, rather than human cognitions. By investigating applications of haptic signals in a realistic context, however, it is expected that cognitive processes that take place during task performance with the tested systems will be similar to those observed in the real world and thus produce comparable behaviour, from which also some insight into human cognition may be inferred.

The proposed application-centred approach is distinctly interdisciplinary as it combines a technological focus on system development with experimental methodologies that were established in the social sciences. Furthermore, since applications of haptic signals, not individual devices, are the focus of this work, a broader perspective in the evaluation of haptic interfaces is encouraged by the chosen approach. The investigations featured in this work will thus further our understanding of haptic human-machine interaction and, at the same time, offer an empirical basis for the system-independent evaluation of haptic signal applications in teleoperation systems, from which transparent guidelines for the design of the haptic human-machine interface may be derived.

Application-centred research and development distinguishes itself from user-centred design, in that task- and user-requirements are not specified prior to system development. Instead, the effectiveness of a developed system is evaluated with regard to specified performance criteria in an applied context. Hence, similar to the prevailing system-centred approach, a sensible application is sought after system development, with the difference that in application-centred design, inter-/and intra-individual differences in observed performance behaviour are taken into account through the use of rigorous user testing, and that the system is tested within a realistic use context. This approach is not intended to provide an alternative to user-centred design, which has proven very effective in product development (Mao, Vredenburg, Smith, & Carey, 2005). Instead, it can build on prevailing system-centred design efforts and process them in a way that will allow for user-centred development of future systems, once they have passed the proto-type stage and a sensible application has been found.

4. Motivation of the present work

In summary, teleoperation systems are human-machine hybrid systems, which allow their users to perform tasks that may be too dangerous or difficult for them to accomplish otherwise. Furthermore, by incorporating the user into the control loop, teleoperation systems are considerably more flexible than automated robotic system and as such represent viable alternatives in cases where automation is not feasible due to practical or financial constraints. Although teleoperation systems are theoretically suitable for a wide range of industrial applications, they are not yet widely employed. A problem, which has been frequently observed and reported for a wide range of systems, is that humans are not very adept at handling the mechanical teleoperator unit in the remote environment. As a result, work performance with these systems tends to be slow and error prone, and there is an increased risk of damaging equipment or material in the remote environment. These performance problems are likely attributable to suboptimal communication between the human and the machine, which may be enhanced with multi-modal human-machine interfaces. In particular, haptic interfaces may have the potential of improving human-machine communication by simulating otherwise missing tactile and kinaesthetic information that humans tend to rely on when performing tasks directly, and by directly guiding the user's control input.

The currently prevailing system-centred approach to the research and development of haptic technology has produced a multitude of haptic interfaces, all of which hold the promise of improving work performance with teleoperation systems. In cases where human performance studies were conducted, these tended to use abstract tasks and unrepresentative samples. Furthermore, the system-centred approach has fostered a fragmented research field in which each study focuses on the effects of one particular system, whilst comparative studies of different systems are virtually non-existent and systematic literature reviews remain rare. As a result, it has become difficult to establish which, if any, applications of haptic signals can be recommended as a measure to improve work performance in an industrial context and how these might be most effectively employed.

In this work, an application-centred approach to the research and development of haptic interfaces proposed, which has the potential of providing a framework that may unite the currently fragmented system-centred field of research and of facilitating interdisciplinary work of engineers and human factors specialists in the future. The motivation driving the present work was three-fold. For one, it was intended to contribute to the small body of reliable, empirical evidence on human performance with haptic interfaces by investigating the effects of four specific haptic applications on measures of work performance. Two applications which aimed at simulating haptic feedback from the remote environment (vibrotactile and force feedback) were investigated, as well as two applications of kinaesthetic guidance (haptic expert demonstrations and haptic assistance functions), whereby not only the theoretical potential of these applications was taken into consideration, but also their present-day practicality. In this context, surveys were also conducted to assess user perceptions, e.g. of usefulness and working comfort with a system. Secondly, based on the results of the four studies presented in this work, as well as evidence gathered from a comprehensive review of published research literature, a comparative study of the four investigated haptic applications in form of a meta-analysis was to be conducted. Thus, it may be ascertained which of these applications hold the most and which have the least potential of increasing accuracy and effectiveness of the human-machine interaction. Thirdly, based on the available research evidence, transparent guidelines for the effective employment of haptic signals were to be derived and compiled in an interactive software package in form of the "HMI (**H**uman-**M**achine **I**nterface) Design Guide".

Chapter II.

The Effect of Haptic Feedback from the Remote Environment on Task Performance

Although haptic displays exist in a multitude of different forms (see Chapter I, Section 3.4.2 for an overview), the present work will focus on haptic feedback devices which are commonly used to signal or simulate forces from the remote environment, and which were found explicitly to have an effect on teleoperator movement coordination and force application. In this chapter, two categories of haptic displays are investigated: force feedback devices and vibrotactile displays. Force feedback devices simulate the forces that would be encountered in direct contact with the remote environment and can stop the motion of a user by applying force via various mechanical solutions (Laycock & Day, 2003). Vibrotactile feedback devices display vibrations to the user in order to alert them to a contact with an object or a surface in the remote environment. Vibratory devices tend to be more cost-effective, lighter, and easier to assemble than most force feedback devices (Cheng, Kazman, & Robinson, 1996). Furthermore, unlike force feedback devices, vibrotactile feedback can be implemented independently of the kinematic control function of the master device (Dennerlein, Howe, & Millman, 1997). For these reasons, they are already widely used in commercially available systems, for instance as rumble features in joysticks and gamepad controls for gaming consoles.

Aiming to ascertain whether simulated tactile feedback from the remote environment improves human task performance with teleoperation systems, vibrotactile and force feedback displays commonly employed in such systems were investigated. Upon review of the literature, it was determined that a direct comparison of these two types of haptic feedback applications in the same context would only be marginally informative. Vibrotactile and force feedback signals are both suited to convey information from the remote environment that would be present in a direct interaction, but lacking during remote operation. However, they are best employed in different situations. Vibratory signals are comparatively good at signalling a sudden state transition (e.g. from movements in free space to an encounter with an object) as they tend to bear no direct physical relation to properties encountered in the remote environment (Iwata, 2008). In contrast, force feedback signals are arguably better suited to inform the user of encountered force intensity as they simulate natural feedback. Consequently, it may be argued that more informative (not to mention fairer) than a direct comparison of vibrotactile and force feedback devices in the same scenario would be separate investigations of these feedback types with respect to plausible alternatives in the respective scenarios.

For this purpose, the effects of vibrotactile signals on task performance are first investigated in a teleoperated pick-and-place task and compared to a visual-only control condition. In a second study, possible advantages of force feedback displays for performance over and above those offered by visual force substitution displays are critically examined with regard to their influence on task completion time and the control of applied forces in a teleoperated abrasion task. Following a summary of experimental findings, the potential of simulating haptic impressions from the remote environment for improving human-machine interaction is discussed and guidelines for the effective use of these two applications of haptic feedback are proposed.

1. An investigation of vibratory signals and their effects on task performance during teleoperated micro-assembly

The effectiveness of vibrotactile devices in improving handling errors, force application and performance speed has hitherto been investigated primarily within the context of human-computer interaction (see Jones & Sarter (2008) for an overview). The idea is that such devices would alert a user in the event of contact with an object, which would lead to shorter reaction times, minimised peak forces and reduced error rates. Various attempts have also been made to use directional vibratory feedback, which, in addition to signalling contact with an object, may also convey information of direction or intensity.

1.1. The effect of directional vibrotactile feedback on force application and performance speed

A number of studies have investigated the effects of directional vibrotactile feedback on task completion time and human force regulation- with varying success. Dennerlein, Shahion & Howe (2000) constructed a vibrotactile sensor and display for a teleoperated submergible robot. According to the subjective impressions of two operators, vibrotactile feedback seemed to reduce the forces that they applied to a surface, and the impression was formed that the task was less difficult to master with their device than without it. Debus, Jang, Dupont & Howe (2002) took a more scientific approach to the investigation of performance benefits as a result of vibrotactile feedback. They have employed a number of disparate vibratory elements that would provide users with a sense of force directionality. They constructed what they termed a multi-channel vibrotactile display, consisting of a cylindrical handle with four embedded vibrating elements driven by piezoelectric beams. The four elements were intended to convey to the user the direction of forces encountered during a teleoperated peg-insertion task. The authors observed that the vibrotactile display reduced peak forces compared to a control condition in which this feedback was not available - but only by nine percent. The authors thought (but did not test) that this effect might be improved if the direction from which the force occurred could be conveyed to the user more precisely, for instance by adding more vibratory elements.

Lathan & Tracey (2002) took a similar approach. They mounted sonar elements to a robot, which could detect obstacles in the vicinity of the robot and conveyed this information to the user via a customised vibrotactile feedback sleeve. Disparate vibrating elements indicated the direction of the obstacle, as participants were instructed to telenavigate the robot through a maze. The results indicated that the vibrotactile feedback reduced the number of collisions with obstacles. There was also a tendency observed for participants to perform faster with the addition of vibrotactile feedback; however, this difference in task completion time was not statistically significant.

Cheng, Kazman & Robinson (1996) tested participants who performed a virtual pick-and-place task with a visual and an acoustic force substitution display and either with or without additional vibrotactile feedback. The authors observed that vibrotactile feedback did reduce task completion times significantly on the first day of trials, but the difference to the control group was diminished by day three. Furthermore, those who received vibrotactile feedback applied overall more pressure on objects compared to those who lacked this type of feedback; a finding which reflects that of an earlier study (Massimino & Sheridan, 1993).

In contrast, Schoonmaker & Cao (2006) found that a similar type of vibrotactile force simulation resulted in significantly lower mean application forces in a simulated minimally-invasive surgery task when applied to the foot. The authors further noted that their vibrotactile feedback improved participants' ability to discriminate between tissues of different consistencies, due to their increased sensitivity to the perception of tissue. The authors themselves, however, questioned the validity of this assumption based on the fact that the average peak forces recorded in both conditions were very similar.

1.2. The effect of non-directional vibrotactile feedback on force application and performance speed

Rather than trying to convey complex information such as force intensity or direction, others have concentrated on using vibrotactile elements simply to alert users in the event of contact in order to reduce excessive forces during teleoperation. Kontarinis & Howe (1995) tested the effects of vibrotactile feedback on reaction time and force control with three participants in a teleoperated task that required participants to pierce a tape with a needle, without exerting excessive force that would damage an underlying rubber layer. The results seem to indicate that vibrotactile feedback significantly decreased peak forces, as well as mean reaction times, defined as the time that passed between piercing the surface and retraction of the needle. In a peg-in-hole task, however, vibratory feedback was not found to affect task completion time, even though the participants stated that it helped them in performing the task. Forces were not measured in this particular task. The authors suggested that vibrations are generally suited to signal transitions between individual task components, e.g. stopping to apply force and starting to retract; however, if a task takes over a second to complete, the achieved reduction in reaction time would not lead to a noticeable reduction of task completion times.

O'Malley and Ambrose (2003) evaluated the use of vibrotactile feedback in the operation of their Robonaut - a humanoid robot designed for space exploration. The objective of their task was to move the robot arm in order to grasp a certain handrail and consequently withdraw it to the starting position without colliding with other handrails in the workspace. This task was conducted with and without vibrotactile feedback in the event of contact. The results indicated that vibrotactile feedback did not significantly improve performance, as it did not reduce peak force magnitudes during contact with the handrail. The authors conjectured that the visual cues in this scenario were sufficient for the task, so that no further improvement could be achieved with the vibrotactile signals. In fact, several studies suggest that haptic feedback in general, and vibrotactile feedback in particular, is most helpful under conditions of poor visibility or ambiguous visual cues (e.g. Bouguila, Ishii, & Sato (2000); Jansson & Öström (2004)). This finding is supported by perception theory, which stipulates that sensory information from different cues interacts in a multiplicative fashion (e.g. Bruno & Cutting (1988); Ernst & Banks (2002)). It follows that under conditions of suboptimal visual feedback, multi-modal displays would have the potential of resolving sensory conflict and subsequent ambiguity, which is particularly relevant in tasks that require accurate judgements of the relative distance between two objects (Eysenck & Keane, 2000).

1.3. The effect of non-directional vibrotactile feedback on performance error and task completion times

The results are equivocal, not the least because explanations for findings are rarely offered. For instance, Akamatsu, MacKenzie & Hasbroucq (1995) found that the time period between reaching a target and selecting it was shorter when participants received additional tactile feedback to the finger pad of the index finger at the moment the target was reached. A subsequent study by the same authors replicated this finding (Akamatsu & MacKenzie, 1996). The experiments also indicated, however, that errors increased with the use of vibrotactile feedback - an effect, which the authors attributed to the participants' unfamiliarity with the experimental task rather than the feedback. Exactly why task unfamiliarity would provoke users to perform their task faster, but more inaccurately with vibrotactile feedback than without it has not been divulged. In contrast, Hasser, Goldenberg, Martin & Rosenberg (1998) reported decreased targeting errors in a point-and-click task with a vibrotactile computer mouse, whereas Forsberg (2008) found no significant performance difference in their point-and-click task between a group that performed this task with added vibrotactile feedback and a control group that only received visual feedback.

Viau, Najm & Chapman (2005) suggested that the effects of vibrotactile feedback on task performance might not only be task-dependent but also influenced by age. Using a computer mouse with vibrotactile feedback, the authors found that tactile feedback improved task completion time and movement accuracy in a task in which participants were instructed to select and drag a word. Specifically, the time taken to select a specific word was reduced by 13%, and the number of errors made while selecting and moving the word was reduced by 23% when vibrotactile feedback was available. However, this effect was only found for younger participants (20-39 years of age), but not for a group of older participants (40-64 years of age). Moreover, participants' performance on two other tasks, menu navigation and selecting & dragging a cell, was not found to be significantly influenced by the presence or absence of vibrotactile feedback from the mouse; nonetheless, a non-significant tendency to deteriorate performance on these tasks was observed with the use of this form of tactile feedback. In accordance with observations made by Valeriani, Ranhi & Giaquinto (2003) several years earlier, the authors suggested that the vibrotactile feedback might have distracted, rather than aided, participants during their performance on these tasks.

In summary, compared to many haptic feedback devices, vibratory displays are cheap and easily implemented into most teleoperation systems. Consequently, there is a vested interest among the proponents of industrial teleoperation systems to establish under which conditions, if at all, vibrotactile feedback can improve task performance with these systems measurably and noticeably. Studies which investigated the effects of vibrotactile feedback on task performance in virtual and teleoperation systems are far from conclusive. Directional vibrotactile displays, which vary the frequency or location of vibration in order to convey information of signal intensity or direction, were not conclusively found to improve any of the investigated aspects of task performance. In contrast, uniform vibration signals that simply inform of the event of contact were reliably found to reduce applied forces under conditions of poor visibility. As of yet, it is unclear, whether these signals would also reduce task completion times or error rate. Yet, one might argue that if vibration signals resolve ambiguity regarding a possible event of contact with the remote environment, not only forces, but also task completion times and error rates should be reduced in a task where this is relevant to the named performance measures.

Since all of the aforementioned performance measures are typically critical in industrial applications of teleoperation systems, experimental user studies are called for in order to systematically investigate in what way these measures are affected by the introduction of vibrotactile feedback in the event of contact. Furthermore, in order to gauge the extent to which the findings of experimental studies may generalise to industrial systems, it is important that the experimental setting, while providing a controlled environment, would reflect realistic task demands.

1.4. Research aim

Hence, aiming to ascertain whether vibrotactile feedback would improve central measures of task performance in a typical industrial teleoperation application, an experimental study was conducted, whereby participants performed a pick-and-place task in a teleoperated micro-assembly setting. The task and its setting were designed specifically to reflect realistic demands as they are typically encountered with this and similar industrial teleoperation systems. A uniform, non-directional vibratory signal was chosen to provide an unambiguous signal of contact with an object in the remote environment. The effects of this type of vibrotactile feedback on task completion time, applied peak forces and pick-and-place success were investigated. Furthermore, in order to acquire some information on the perception of vibrotactile feedback, participants were asked a series of general questions pertaining to equipment operation and perceived effects of settings.

1.5. Research hypotheses

Specifically, the following predictions were made:

H1: Studies suggest that participants are able to make contact with a specific target on a surface more accurately if they receive haptic feedback from the surface rather than just an ambiguous 2-D view of the surface (Bouguila, Ishii, & Sato (2000); Jansson & Öström (2004)). It is therefore believed that participants will be able to pick and place their microchip more successfully, if they receive a vibratory signal in the event of contact. Conversely, pick-and-place performance will deteriorate with the omission of this signal.

H2: Vibrotactile feedback is believed to be suited to signal the transition between one step of the task and the other (Kontarinis & Howe, 1995). It was therefore predicted that vibrotactile feedback will allow participants to perform their task more quickly, since the arm can be quickly lowered to the surface until the vibrotactile signal occurs. Without this transition signal, participants need to check continuously the position of the arm in relation to the position of the table and therefore will require comparatively more time for the task.

H3: Since vibrotactile signals are believed to signal the user to retract the arm unambiguously (Kontarinis & Howe, 1995), it is further predicted that participants will also be more likely to apply less pressure to the table if they receive a vibratory signal in the event of contact than if this feedback is missing.

1.6. Method

1.6.1. Participants

An opportunity sample of 24 male and eight female participants (N = 32) took part in the experiment (M_{age} = 24 yrs., SD_{age} = 3.70 yrs.), all of whom were right-handed.

1.6.2. Experimental apparatus

The experiment was conducted in collaboration with the Institute for Machine Tools and Industrial Management (*iwb*) of the Technische Universität München, which provided hardware and software, and implemented the teleoperation system.

Teleoperator Environment

Figure 13 shows a picture of the teleoperator used in this study. One of the arms of the teleoperator was equipped with a pneumatic two-finger gripper, with which microchips (size 1.3 mm x 1.3 mm x 0.5 mm) could be grasped and transported. In the clamping parts of the gripper, a force sensor was installed which detected contact with the gripper as well as unintended collisions. Detected collisions were conveyed to the user in the form of vibrations of the Saitek Cyborg Evo Force joystick, with which the user controlled the teleoperator arm. The arm was equipped with a linear drive which moved the arm into z-direction with an accuracy of 1 μm. Flexibility of the teleoperator was ensured with a planar table (x- and y-axis), which provided an accuracy of 1 μm for precise micro-positioning combined with a high action radius (up to 50 cm x 80 cm).

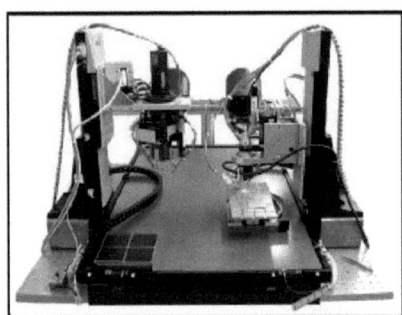

Figure 13. The micro-assembly teleoperator setup.
Source: Reiter, et al. (2008).

Operator environment

The operator workplace was comprised of a standard gaming joystick (Saitek Cyborg Evo Force) as haptic input/output device and a monitor for the visual feedback (see Figure 14). According to Mattos & Caldwell (2009), the joystick featured two rotation axes with force feedback capability (pitch and row) and a spring-loaded yaw axis. Further technical details of the experimental set-up are described in Reiter, Nitsch, Reinhart & Färber (2008).

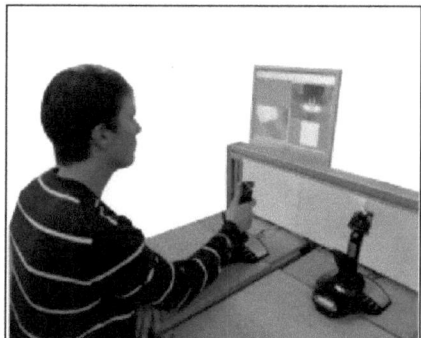

Figure 14. Operator workplace.

In order to move the teleoperator's gripper arm up and down, a button needed to be pressed and the joystick had to be moved in y-direction. The gripper itself opened and closed if another button was pressed.

Two video cameras provided the user with visual feedback from the remote environment via live video stream: an integrated telecentric scaling camera showed a magnified view vertically downwards directly over the gripper, while the other camera displayed an overview of the task environment from a flat angle against the horizontal line (see Figure 15).

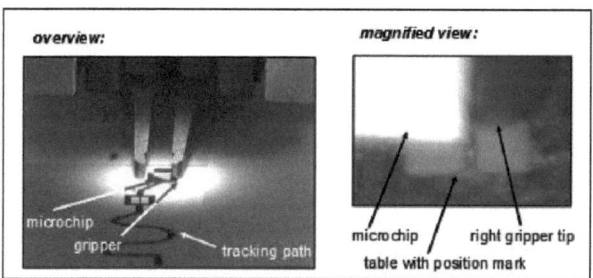

Figure 15. Visual feedback from the remote environment.
Source: Reiter, et al. (2008).

Great care was taken to achieve a realistic set-up as it would be encountered in industry, rather than contriving a situation in which the view would be obscured or the quality of the video feedback degraded. Hence, the camera angles and positions were chosen to allow for the best possible visual feedback necessary in order to perform the task well. Nevertheless, during pilot testing it was established that ambiguity regarding a possible contact between gripper and table could not be fully resolved with visual feedback alone, thus rendering this task particularly suitable for the investigation of the effects of tactile feedback in the event of realistic visual ambiguity.

1.6.3. Experimental design

The aim of this study was to test whether users performed better in a pick-and-place task if they received vibrotactile feedback rather than just video feedback from the remote environment. For this purpose, a 2 (haptic feedback type) x 2 (task) within-subjects design was utilised. Haptic feedback was manipulated on two levels (vibrotactile feedback on/off), as was the task to be performed (picking/placing). The time it took participants to pick and place the chip (sec.) was measured with a stopwatch, based on the video analysis of each trial. The forces exerted on the table in the event of contact were also measured (N). Successful grasping of the chip and successful placing of the chip within the confines of the drawn target boxes were assessed as dichotomous measures (successful/unsuccessful). Pick-and-place performance was chosen as a measure of the suitability of vibrotactile feedback in signalling transitions between task components, as the key to successful performance lies in the accurate positioning of the gripper above the planar table. That is, if the gripper's position is too high or if it is pressed onto the surface too firmly, the microchip will not be picked up or placed correctly. Hence, the success rate of this task is greatest if the chip is grasped or placed as soon as contact is first established with the table. In the vibrotactile feedback condition, this is signalled by the onset of vibrations. In the control condition, the user has to rely on visual cues from the video feedback in order to judge whether the gripper is touching the table's surface. The design was balanced and the conditions were presented in a systematically randomised order to counterbalance sequence and fatigue effects. The video feedback was available to all participants at all times, and the two different camera views were always displayed in the exact same position on-screen. The entire experiment took place in a sound-attenuated room to ensure that participants were not unduly affected by extraneous influences.

1.6.4. Procedure

With standardised instructions, participants were asked to pick up a microchip, transport it to the next box drawn on the track, place it as accurately as possible within the confines of that box, retract the teleoperator arm again, and move to the pre-defined end position. Participants were told to perform this entire task as accurately and as quickly as possible. The task was repeated eight times with and eight times without vibrotactile feedback. Half of all participants started with vibrotactile feedback, the other started without it. Participants were informed that on some occasions, they might experience vibrations when coming into contact with the surface. They were not told that these vibrations served a particular purpose, nor were they told to respond to these vibrations in a particular way. The decision to withhold this information was made in order to avoid an expectation bias by instilling an expectation in the user that vibrotactile signals should have an effect, as well as to avoid that participants would apply excessive forces during trials without vibrotactile feedback, expecting to receive a signal. After all trials had been completed, a survey was conducted; asking participants if they noticed the vibrotactile feedback and whether they believed the vibrotactile feedback improved their performance. Care was taken to formulate non-leading questions and to ensure conditions of anonymity. Afterwards, participants were debriefed regarding the purpose of the experiment.

1.7. Results

First, it was tested whether the assumptions for parametric tests were met. Appendix A offers definition of these assumptions and the methods used for validation.

Data with z-values[1] of greater than ±3.29 were considered outliers and consequently excluded from further analysis. In cases in which the assumptions were violated, outliers were identified via boxplots and corrections or non-parametric tests were subsequently applied. Even though the stipulated hypotheses are directed based on the findings of previous studies, the empirical base was not considered sufficient to justify one-sided statistical tests. Hence, inferential statistics in this and the following studies always report two-sided (exact) α-probability[2] values.

---

[1] Z-values are observed scores expressed as standard deviations of a normal distribution. The z-distribution has always a mean of 0 and a standard deviation of 1. Since the probability of a z-value of ±3.29 occurring in a normal distribution is approximately 0.1%, z-scores with values >±3.29 are considered sufficiently unlikely to belong to the same population as the other observed values and may hence be considered outliers (Field, 2009, p 26).

[2] In short, α-probability values indicate the probability that a stipulated statistical model (i.e. the experimental hypothesis) would describe the observed data as well as it does, by chance. If it is sufficiently unlikely that a comparable model fit would also apply to a chance distribution, the null hypothesis is rejected and the experimental hypothesis is accepted. By convention, the accepted two-sided α-level lies at $p<.05$, meaning that there is a chance of less than five percent that the observed data could occur by chance rather than being caused by the experimental manipulation. If the direction of the effect is predicted, the accepted one-sided α-level can be upwardly adjusted (Field, 2009).

1.7.1. Pick-and-place performance

H1: Studies suggest that participants are able to make contact with a specific target on a surface more accurately if they receive haptic feedback from the surface rather than just an ambiguous 2-D view of the surface (Bouguila, Ishii, & Sato (2000); Jansson & Öström (2004)). It is therefore believed that participants will be able to pick and place their microchip more successfully, if they receive a vibratory signal in the event of contact. Conversely, pick-and-place performance will deteriorate with the omission of this signal.

The data revealed that, on average, the microchip was picked up successfully in 94.53% and 95.70% of trials with and without vibrotactile feedback, respectively. Of those trials in which the chip was successfully picked up, it was placed correctly in 67.69% of trials when no vibrotactile feedback was implemented. In contrast, when vibrotactile feedback was added, it was only placed correctly in 59.61% of trials (see Figure 16).

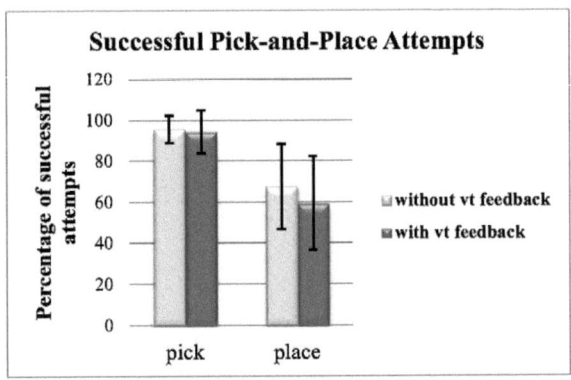

Figure 16. Respective mean percentages of successful pick and place movements of all attempted trials. Error bars indicate standard deviations.

Due to a violation of parametric assumptions, non-parametric statistics were applied to this data set. A direct comparison of the successful pick and place attempts with both types of feedback using Wilcoxon signed-rank tests[3] showed no significant difference in picking ($z_{picking}$ = -.70, p =.49) or placing ($z_{placing}$ = -1.24, p =.22), suggesting that overall, adding vibrotactile feedback did not significantly affect picking nor placing performance. Thus, the data show no support for hypothesis H1.

1.7.2. Task completion time

---

[3] Wilcoxon signed-rank tests is a non-parametric test that compares the ranked difference scores of two dependent samples, and which was shown to be robust to small sample sizes. Hereby, significance values are calculated based on the smaller of the two respective sums of positive and negative differences (Howitt & Cramer, 2003).

H2: Vibrotactile feedback is believed to be suited to signal the transition between one step of the task and the other (Kontarinis & Howe, 1995). It was therefore predicted that vibrotactile feedback will allow participants to perform their task more quickly, since the arm can be quickly lowered to the surface until the vibrotactile signal occurs. Without this transition signal, participants need to check continuously the position of the arm in relation to the position of the table and therefore will require comparatively more time for the task.

The mean task completion times indicated slightly shorter task completion times for trials conducted with vibrotactile feedback (see Figure 17).

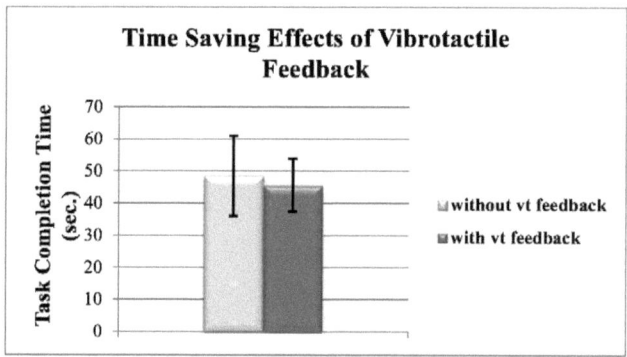

Figure 17. Mean task completion times (sec.) and their standard deviations for trials conducted with and without vibrotactile feedback.

However, repeated-measures t-tests[4] indicated no significant difference in task completion times between the two conditions (t(31) = 1.34, p = .19), thus showing no support for hypothesis H2.

1.7.3. Excessive force application

H3: Since vibrotactile signals are believed to signal the user unambiguously to retract the arm (Kontarinis & Howe, 1995), it is further predicted that participants will also be more likely to apply less pressure to the table if they receive a vibratory signal in the event of contact than if this feedback is missing.

Due to temporary equipment failure, force measurements were available for 15 participants. To test the hypothesis, the peak forces of all eight trials with vibrotactile feedback were compared to the eight trials performed without this type of feedback (see Figure 18).

---

[4] The repeated-measures (or dependent) t-test is a parametric test which compares the mean scores of two related samples and tests whether these means could be samples of the same population. The reported significance values indicate the likelihood with which the observed difference between mean scores could occur, assuming that the null hypothesis is true. The probability is calculated based on the t-distribution (Howitt & Cramer, 2003).

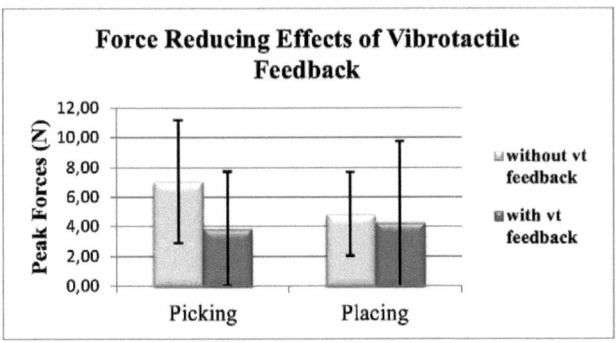

Figure 18. Mean peak forces (N) and their standard deviations for picking and placing acts with and without vibrotactile feedback.

The mean peak forces indicate that lower peak forces were observed with vibrotactile feedback than without it. During picking, the difference in peak forces is particularly noticeable. However, much larger standard deviations were found for trials conducted with vibrotactile feedback than for those without it, indicating that the peak forces observed in this condition varied greatly between individuals. Due to the small sample size, inferential statistical analyses were conducted with non-parametric tests, as recommended by Field (2009). Wilcoxon signed-rank tests confirmed that when participants picked the microchip up, significantly larger peak forces were recorded for trials without vibrotactile feedback, than for those with this additional feedback option ($z_{picking}$ = -2.84, p<.01, r = -.73[5]). In fact, the effect strength indicates a very strong impact of feedback on peak forces. However, for the act of placing, peak forces did not significantly differ between trials ($z_{placing}$ = -1.25, p = .21). Since the hypothesis did not differentiate between types of contact situations, one might argue that hypothesis H3 is partially supported by the experimental evidence.

1.7.4. Survey results

Survey data were available for 31 participants. Participants were asked whether they noticed vibrations during some of the trials when they performed their task and, if they did, whether they thought the vibrations affected their performance in any way.

Only 21 out of 31 participants (67.74%) stated that they had noticed the vibrations. Of those 21 participants who did notice the signals, six (28.57%) stated that they felt adding vibrotactile feedback made no difference with regard to their task performance. Seven participants (33.33%) claimed that

---

[5] In this context, Pearson's correlation coefficent r is used as a measure of effect size. Cohen (1988) proposed that a value of r =.10 signifies a small effect, a value of r = .30 a medium effect. A value of r = .50 denotes a large effect, accounting for 25% of variance in the observed scores.

the vibrations made their task more difficult to perform, whereas eight (38.10%) believed the vibrotactile feedback aided their performance.

1.8. Discussion

The assumption was investigated that adding vibrotactile feedback would improve task performance in a pick-and-place task, as it signals contact with the table unambiguously, whereas a two-dimensional camera view would have the potential of introducing ambiguity associated with (a lack of) visual cues of depth. In accordance with expectations was the finding that vibrotactile feedback significantly reduced peak forces when picking up the chip, although, rather unexpectedly, there was no significant difference during chip placing. The latter findings may be explained by the fact that users did not always establish contact with the table when placing the chip but on some occasions dropped it instead. In contrast, the user had no choice but to come into contact with the table in order to pick up the chip. Since the application of pressure was not necessary for placing the chip, the risk of applying excessive force is much lower, regardless of the feedback received. This would also explain the much lower success rate for placing than for picking, since both acts can only be performed successfully if the gripper is in contact with the table. Overall, picking and placing performance was neither significantly better nor worse when vibrotactile feedback was added, as the users did not perform significantly faster nor more accurately when vibrations signalled the contact with the table. The results of this study thus imply that although users may be motivated to apply less pressure to a surface when encountering vibrotactile signals, they do not necessarily react more quickly nor more appropriately to a vibrotactile signal compared to a visually ambiguous signal of contact.

It is conceivable that vibrotactile feedback might be more effective in improving measures of task performance with repeated conscious exposure during training. Supportive of this notion is the fact that almost a third of people had stated not to have noticed the vibrations, even though they had been informed of their possible occurrence beforehand. There are several possible explanations for this finding. One would be a response bias. In fact, when users were informed of the purpose of the vibrations in the debriefing phase, some of them stated that they had attributed these to a glitch in the system. Thus, it is conceivable that participants dismissed the vibrations as irrelevant to the task, or that participants did not wish to embarrass the experimenter by stating their system was faulty. Alternatively, although the vibration frequency well surpassed the perception threshold, the vibrations may have lacked the intensity to capture the users' attention while they concentrated on performing their task. If this assumption is correct, it is conceivable that vibrations with greater amplitude might have a greater impact on task performance measures. Rather than dismissing the effect of vibratory signals on task performance measures of time and error rate as negligible, it is therefore suggested that these questions should be subject of further investigations before more substantive conclusions can be drawn.

The survey data acquired in the present study further suggest that the type of haptic feedback featured in this experimental study was not as accepted by users as it would have been expected based on previous surveys (e.g. Gill & Ruddle, (1998)). Reason for this lack of acceptance may be that the vibrotactile feedback used in the present study does not constitute a realistic representation of the real world stimulus, since such vibrations would not occur during direct manipulation. This points to the importance of providing feedback that is either analogous to the "real world" stimulus or widely accepted in order for it to improve performance. To ensure that the provided vibrotactile signal can be interpreted intuitively and meaningfully as a state of transition that requires action, it may be necessary to condition a particular response to the vibratory stimulus first. Yet, with around 50% of variance in peak force during picking accounted for by the addition of vibrations, the results of this study suggest that even without prior conditioning, a considerable benefit of employing vibrotactile feedback in reducing the risk of damaged equipment or material due to excessive force application in the event of contact.

Overall, the available research evidence suggests that vibratory feedback, if at all, is only likely to improve some aspects of task performance, for some tasks, some of the time. Yet, the present study shows rather conclusively that particularly in industrial applications, in which rather fragile material or expensive equipment can be damaged easily through the imprecise application of forces, vibratory feedback offers a cost-effective and easily implemented means of reducing the risk of damage caused by excessive forces.

2. An investigation of force feedback displays and their effect on task performance during a teleoperated abrasion task

For some tasks, it may not suffice to be informed of a contact, as it may also be necessary to receive information of the *intensity* of encountered forces for performance of some tasks, such as abrasion, palpation or tissue puncturing. As it was previously discussed in Part II, Section 1.3, some vibrotactile devices have been developed which aim to increase the vibration rate or intensity congruously with increasing forces encountered in the remote environment, thus conveying the user an approximate idea of force intensity (e.g. Cheng, Kazman, & Robinson (1996); Schoonmaker & Cao (2006), Chatterjee, Chaubey, Martin, & Thakor (2008)). While the research evidence on the effectiveness of vibrotactile displays in conveying information of force intensity is rather ambiguous (see p. 39 for a discussion), there is considerable empirical evidence suggesting that force feedback displays assist the user's regulation of forces applied to surfaces or objects in the remote environment.

2.1. The effect of force feedback on force application

Studies on the effects of force feedback displays on human task performance with teleoperation and virtual systems abound, with most being conducted within the context of teleoperated surgery. More than two decades ago, Draper, Hemdon & Moore (1987) and Kim (1992) had reported lower peak forces and lower cumulative forces, respectively, during surgery-related tasks and attributed these to the use of force feedback devices. Kazi (2001) even reported reductions of maximum forces of up to 40% with the use of force feedback during various teleoperated surgery tasks such as tissue puncturing and catheter insertion. Later on, Wagner, Howe & Stylopoulos (2002) examined the effects of force feedback during a blunt dissection task, in which participants were asked to expose an artery in a synthetic model while viewing the teleoperator site via laparoscope video feedback. Participants performed the dissection task without force feedback, with force feedback scaled down to 75% of the original forces measured at the teleoperator site, and with a force feedback scaling of 150%. The authors reported that compared to the control condition in which participants did not receive force feedback from the remote environment, peak force values and force variance were significantly reduced when force feedback was added, with further reductions observed with increasing force scaling.

A similar study was conducted several years later by Mayer, Nagy, Knoll, et al. (2007), who tested their ARAMIS system (Autonomous Robot Assisted Minimally Invasive Surgery) with surgeons in a number of experiments from which they concluded that force feedback improved various aspects of surgeons' work performance with the system, in particular those mandating precise control of applied forces. Specifically, they compared task performance in a control condition without force feedback to a condition in which measured forces were fed back to the user directly and to a condition in which force feedback was amplified by a factor of two. Compared to the control condition, which lacked haptic feedback from the remote environment, amplified force feedback was found to reduce significantly applied forces during knot tying as well as during a palpating task, which aimed at the detection of arteriosclerosis. Moreover, it allowed participants to estimate the point of suture material breakage significantly more precisely. The authors also reported that the presence of haptic feedback was found to reduce surgeons' fatigue significantly, as indicated by the critical flicker fusion frequency. However, from their account it is unclear whether this finding extends to both levels of force feedback tested, or only applies to the amplified feedback. It also appears that non-amplified force feedback exerted no significant effect on any of the measures when compared to task performance in the control condition. Whether this lack of effect argues for the amplification of forces in general or indicates that the effectiveness of force feedback depends on the absolute intensity of forces encountered, is debatable.

2.2. The effect of force feedback on other performance aspects

Other aspects of task performance have also been found to improve with force feedback displays, although the empirical evidence to this issue is rather equivocal and suggests that performance-enhancing aspects of force feedback are task-dependent. For example, Draper, Hemdon & Moore (1987) reported lower error rates with force feedback devices in assembly tasks, while Kitagawa, Okamura, Bethea, et al. (2002) demonstrated improved knot-tying performance during robot-assisted surgery. In contrast, Wagner, et al. (2002) stated that the length of artery dissected and the tissue area affected per cm of artery dissected within a space of five minutes were not significantly affected by the presence of force feedback. Similarly, Mayer, et al. (2007) showed that scaled and non-scaled forces had no significant effect on task completion times, the number of knots tied within a certain time frame or arteriosclerosis detection errors.

A number of studies found force feedback to increase the speed with which people were able to perform tasks of varying complexity successfully, i.e. performance efficiency. For instance, a study by Massimino & Sheridan (1992) demonstrated completion time improvements when using force feedback under differing degrees of visual degradation. Simulating an assembly task without visual impairment, Gupta, Sheridan & Whitney (1997) confirmed shorter task completion times in a peg-in-hole scenario, when participants could feel the objects with thumb and index finger, using two PHANTOM force feedback devices. Similarly, Cao, Zhou, Jones & Schwaitzberg (2007) investigated whether surgeons would perform a pick-and-place task faster with a laparoscopic simulator with haptic feedback. They determined that, on average, haptic feedback decreased task completion times by 37%. Arsenault & Ware (2000) reported a statistically significant reduction of 12% in the inter-tap interval in a Fitts[6] task with haptic feedback. Sallnäs (2000) investigated collaborative task performance and found that participants managed to construct a tower of virtual cubes significantly faster when they received haptic feedback from the environment and, consequently, also other users, thus replicating an effect observed by Ishii, Nakata & Sato (1994).

However, not all studies pointed to increased efficiency with the use of force feedback. Hurmuzlu, Ephanov & Stoianovici (1998) found that users did not perform significantly faster when they received force feedback during a drawing task on a pressure-sensitive tablet. Similarly, Oakley, McGee, Brewster & Gray (2000) reported no performance time improvements in interactions with a GUI when adding force feedback. Deml, Ortmaier & Seibold (2005) even reported slightly slower times (by 9.4%) for a rather complex, robot-assisted minimally-invasive surgical task when force feedback was added; however, this difference was not statistically significant. In each of these studies, performance accuracy was found to be improved significantly by the addition of force feedback.

Overall, the empirical evidence indicates that while force feedback is likely to reduce peak forces, increase force application accuracy and decrease performance errors, it does not necessarily improve the speed of the human-machine interaction. Although, to the author's knowledge, it has not been explicitly stated nor investigated, it seems most likely that force feedback would primarily affect human force regulation, which may then act as a mediator of task completion times. It would therefore most likely depend on the performance characteristics of the task, whether or not force feedback reduces completion times. That is, if force regulation accuracy is related to performance speed for a particular task positively, force feedback will be likely to reduce forces and increase speed. If there is a speed-accuracy trade-off, force feedback will be likely to increase regulation accuracy at the cost of speed. On the other hand, if applied forces and task completion times are unrelated, the use of force feedback will still reduce forces, but not task completion times.

2.3. Force sensory substitution displays

As a more viable alternative to less effective vibrotactile displays and expensive force feedback devices which are oftentimes difficult to implement in industrial teleoperation systems, it has been suggested to convey haptic information through other sensory channels, for example, by providing visual displays which inform the user of applied forces (e.g. Kitagawa, Dokko, Okamura, & Yuh

---

[6] A Fitts task is designed after Fitts' (1954) law which stipulates a formal relationship of speed/accuracy trade-offs in quick, aimed movements. Specifically, it states that a logarithmic function of the spatial relative error in such movements describes the time to move and point to a target of a certain width at a specified distance. A point-and-click task would be a typical example of a Fitts task.

(2005); Tavakoli, Patel & Moallem (2005)). Visual-to-tactile sensory substitution, in which case visual information is conveyed via tactile stimuli (e.g. braille), has been examined in great depth. In contrast, much less is known regarding the effectiveness, i.e. the success in accomplishing performance goals, and efficiency, i.e. the speediness in accomplishing performance goals successfully, of tactile-to-visual sensory substitution.

2.3.1. Theory on the effects of force feedback and force sensory substitution displays on task performance

Human beings are accustomed to perceive the world primarily through their eyes (Mauter & Katzki, 2003). It is therefore hardly surprising that the human visual system is highly sophisticated and quite adept at integrating different pieces of information into coherent percepts (Wickens A. , 2000). Despite the highly developed capabilities of the human visual system, research suggests that haptic displays may be superior to visual displays of haptic information in enhancing performance with teleoperation systems. In fact, two theories rooted in cognition research seem to point to this conclusion: Wickens' (1984) resource theory and the stimulus-response compatibility theorem.

For one, when speculating about potential beneficial effects of haptic displays on work performance, studies on haptic displays tend to cite Wickens' (1984) resource theory for multiple task performance. Reviewing the empirical evidence on dual-task performance available at the time, Wickens concluded that when a user performs two tasks at the same time, performance on one or both of these tasks may suffer if they share a stimulus modality, draw on the same stages of processing (input, internal processing and output), and/or rely on related memory codes (Eysenck & Keane, 2000). Since work performance with teleoperation systems in applied settings almost always requires the simultaneous performance of two or more tasks (e.g. correctly positioning the manipulator in three-dimensional space and applying an appropriate amount of force), this theory would also seem to apply to the teleoperation context. For example, when applied to a teleoperation context, the theory would imply that if the user is required to pay attention to the positioning of the teleoperator and applied forces simultaneously, performance on one or both of these tasks might suffer if the user has to rely on visual input for both tasks simultaneously. On the other hand, there should be no interference if only one of these tasks (positioning) taps the visual modality, while the other (force application) draws on a different sensory modality.

Secondly, although auditory displays are often recommended in the design of machines in order to reduce visual demands (e.g. Brown, Newsome, & Glinert (1989); Liu (2001); Szalma, et al. (2004)), research suggests that tactile displays in particular may hold advantages over the use of other sensory modalities for some tasks. For example, ample empirical evidence has confirmed that task performance is affected by stimulus-response compatibility (SRC). Whilst its effects are empirically well documented, varying definitions have been proposed for this concept (see Kornblum, Hasbroucq, & Osman (1990) for a comprehensive review of the concept). Sanders (1980) defined SRC as "the degree of natural or over-learned relations between signal and responses" (p. 339), others ascribed it to the sheer physical similarity between stimulus and response sets (Fitts & Seeger (1953); Rosch (1978)). In particular, increased SRC was found to be causally related to reduced reaction times (Blackman (1975); Schwartz, Pomerantz, & Egeth (1977)) and reduced error rates (Gehring, Gratton, Coles, & Donching, 1992). Applied to teleoperation, this concept would suggest that users might adjust their forces more quickly, if the input signal (stimulus) shared the same, rather than a different, modality to the output signal (response). Thus, according to the SRC concept, a tactile display of forces would be less likely to produce excessive forces on part of the user, than a visual display of forces would be, as the user would be able to retract the manipulator more quickly in reaction to a tactile signal than to a visual signal. Moreover, by increasing the effectiveness of the human-machine interaction, it would seem likely that it might also reduce overall performance times for some tasks.

To summarise, while Wickens' (1984) theory of multiple task performance would suggest that task performance improves if the signal *input* for two different tasks do not share the same sensory modality, SRC stipulates that task performance improves if *input and output* for one task share the same modality. Thus, it may be deduced that performance with a teleoperation system will increase if two input signals are presented to two different rather than the same sensory modalities, and if the response to this task utilises the same rather than a different modality to the input signal. Since both of these premises can be fulfilled with haptic interfaces, it stands to reason that haptic interfaces, in particular those that provide force feedback, would improve task performance with teleoperation systems compared to systems that only utilise visual or visuo-auditory feedback.

2.3.2. Empirical findings on the effects of force feedback and force sensory substitution displays on task performance

In the vast majority of experimental studies on the value of force feedback during teleoperated tasks, performance with haptic feedback is always compared to a condition, in which the force information could only be deducted implicitly through the observation of visual cues. Yet, auditory displays and visual force feedback substitution displays would present cost-effective options, with most implemented much more easily compared to presently available force feedback devices. Although studies on sensory substitution of haptic information have increased in recent years (e.g.

Massimino & Sheridan (1992); Kitagawa, Dokko, Okamura, & Yuh (2005); Bethea, et al. (2004)), only few studies directly compared task performance with force feedback displays to that conducted with force substitution displays.

One of the few studies investigating this question was performed by Massimino & Sheridan (1992) who concluded that auditory and vibrotactile force displays compared favourably to force feedback devices in teleoperated peg-in-hole tasks of varying complexity. Almost a decade later, Williams, Chen & Seaton (2002) presented evidence suggesting that substitution displays may not only lead to comparable success in task performance, but may also produce similar effects on user force control. The authors investigated the effects of force feedback on teleoperator performance in a space operations drill task with four different force display options: stereo video feedback only, visual force overlay, haptic (kinaesthetic) force reflection and a combination of these three options. Although task completion times did not significantly differ between any of these conditions, the visual force display reduced peak forces and torques by an average of 23% compared to the control condition, which lacked any force information. Kinaesthetic force feedback proved to be superior still with an average maximum force and torque reduction of 43% and 27%, respectively, compared to the control condition. The combination of the visual force overlay and the kinaesthetic force reflection showed no further improvement to that of the kinaesthetic display option. Unfortunately, no inferential statistics were reported for this study, so it remains unclear to what extent these findings might generalise beyond this study. Moreover, during drilling, the user can focus on applying a constant force for a certain period of time. Many industrial tasks, however, require more dynamic movements, during which the operator's focus is split between the application of force and the correct positioning of the device. A question yet unanswered is, therefore, whether visual force displays are a viable option to haptic force displays in dynamic tasks which would require the user's visual focus to shift continuously between the display and the teleoperator's position.

2.4. Research aim

Aiming to ascertain whether force feedback would improve force regulation accuracy in a task that requires the operator to apply a constant pressure to an object during a dynamic positioning task, a study was conducted in which users were instructed to perform teleoperated abrasion of a non-compliant surface with an industrial robot. In this context, the effect of force feedback on performance speed was also investigated. Previous studies suggest that force feedback will only affect task completion times indirectly by supporting the user in their force application. The results of this study could further serve to explore the relationship between force feedback, force regulation and performance speed.

Since visual substitution displays for haptic information are cheap and easily implemented, and considering the large forces that can occur during remote manipulation, an experiment incorporating these displays would seem more realistic and plausible than a scenario in which users received no visual information of the forces that they apply. Hence, rather than comparing performance with direct force feedback and force substitution to a control condition lacking any type of force information, the goal of this experiment was to ascertain whether force feedback would provide additional assistance over and above that offered by the visualisation of applied forces during such a task. In this context, it was also aimed to assess user perceptions of functionality and usefulness of the provided force feedback.

2.5. Research hypotheses

The following research hypotheses were tested:

H1: User force regulation will benefit from force feedback from the remote environment (Draper, Hemdon, & Moore (1987); Wagner, Howe, & Stylopoulos (2002)).

> *H1a: Maximum applied forces will be significantly reduced with the presence of force feedback compared to performance in its absence.*
> *H1b: Greater variations in applied forces will occur without the use of force feedback from the remote environment compared to variations in forces applied with the support of force feedback.*
> *H1c: The users' regulation of applied forces will be more accurate to the task requirements when force feedback is present than when it is absent.*

H2: The effect of force feedback on task performance speed will depend on the performance characteristics of the task (e.g. Massimino & Sheridan (1992); Cao, Zhou, Jones, & Schwaitzberg (2007)). If forces and task completion times correlate significantly, force feedback will increase accuracy, and either increase or decrease completion time, depending on the sign of the correlation. If applied forces and task completion times do not significantly correlate, force feedback will significantly affect force regulation but not task completion time.

2.6. Method

2.6.1. Participants

An opportunity sample of 30 male and four female participants (M_{age} = 26.4 yrs., SD_{age} = 3.19 yrs.) took part in the study. Although four participants stated to write with their left hand, all participants used their right hand to operate a computer mouse and intuitively grasped the joystick with their right hand. Since basic motor skills such as those involved in moving a joystick are, once learned, not easily forgotten (Wickens A. , 2000), the accumulated lifetime joystick experience rather than average experience over a limited time period was assessed. Seven participants stated to have no prior experience in the use of a joystick (novice), 15 participants had experience of 1-50 hours over their lifetimes (beginner), whereas twelve participants had experience of more than 50 hours of joystick operation (expert). Overall, participants had an average accumulated joystick experience of $M_{experience}$ = 154 hrs. ($SD_{experience}$ = 381 hrs.). A pre-screening of participants ensured that none were impaired in their motor ability.

2.6.2. Experimental apparatus

As before, the experiment described in the following was conducted in collaboration with the Institute for Machine Tools and Industrial Management (*iwb*) of the Technische Universität München, which provided hardware and software, and implemented the teleoperation system.

Teleoperator Environment

An industrial KUKA KR100 robot was used as a teleoperator, a 6-DOF articulated industrial robot with a payload of 100 kg. The robot has a controller with a real time communication interface, the KUKA Ethernet Remote Sensor Interface (RSI). The RSI is used to connect the teleoperator with the bilateral central controller. The exchanged data are transmitted via the Ethernet TCP/IP protocol as XML (Extensible Markup Language) strings. The cyclical data transmission from the robot controller to the central controller is in the interpolation cycle of four milliseconds. This interface allows a direct intervention in the path planning of the robot during motion. The force feedback used in this experiment was a model-based force feedback, designed to scale down the measured forces to ensure stability of the teleoperation system with the low-inertia haptic joystick and the high-inertia teleoperator.

In order to simulate the forces occurring during industrial abrasion tasks, a ball bearing is employed as manipulator (see Figure 19).

Figure 19. Kuka KR100 Robot with ball bearing end-effector.

It can support both radial (perpendicular to the shaft) and axial (parallel to the shaft) loads. Moreover, it reduces the rotational friction between the object surface and the robot's tool and allows for the measurement of the contact forces of the tool perpendicular to the object surface. The force direction is then used to generate the model-based force signals to be fed back to the human operator.

Operator Environment

A 2-DOF force feedback joystick by Immersion Corp. is used in this setup as an input device. This joystick can display forces up to 8.9 N in both directions. Participants were seated in a sound-attenuated room and received visual feedback from the teleoperator via a live video feed.

2.6.3. Experimental design

The task objective was to simulate a dynamic abrasion task by applying a constant pressure of between 300 N and 500 N to a solid steel block whilst moving a certain distance. The specified force range had been established as sensible and plausible during extensive pilot testing. A repeated-measures, within-subjects design was used, with force feedback manipulated on two levels (on/off). A vertical bar graph, which visualized the applied forces, was displayed to the user at all times. With measured forces in excess of 500 N, the bar was displayed in red, turned yellow with forces lower than 300 N, and green if the forces stayed inside the desired range (see Figure 20). The bar graph was always situated to the left of the video feedback of the remote environment.

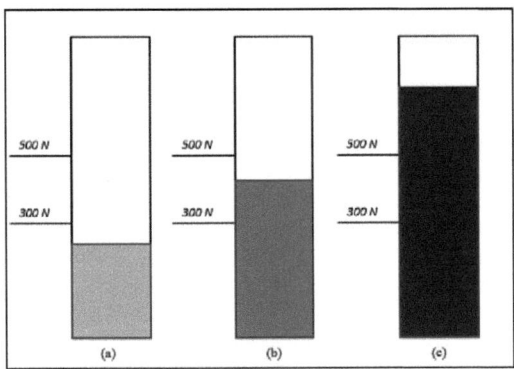

Figure 20. Depiction of the visual force substitution display. In case (a), the user applies too little force, in case (c) too much. Case (b) indicates the optimum amount of applied force.

Participants were randomly assigned to one of two conditions: half of all participants started with force feedback from the remote environment, the other half started without force feedback. Measured were task completion time and the applied forces. Upon conclusion of the experiment, participants were asked a number of questions regarding the quality of the perceived feedback and its perceived usefulness.

2.6.4. Procedure

After giving informed consent, participants were instructed to lower the end-effector of the teleoperator until contact with a solid steel bar was made and then move the effector to a pre-defined goal position to the right as quickly as possible, whilst applying a constant force of between 300 N to 500 N. To ensure that all participants had comparable levels of experience with this task, this task was first practiced three times, before it was repeated another three times for the experimental runs. This procedure was then repeated either with or without added force feedback. The three repetitions were primarily intended to obtain data that are more representative of the true performance score, rather than for the purpose of investigating practice effects.

2.7. Results

Based on the central limit theorem, the data were assumed to be normally distributed (Field, 2009). The data were screened for outliers and data with z-scores of ≥3.29 were replaced with variable means[7]. Further assumptions of parametric data were met; therefore, parametric tests have been applied for further data analysis (see Appendix A).

2.7.1. Excessive force application

H1: User force regulation will benefit from force feedback from the remote environment (Draper, Hemdon, & Moore (1987); Wagner, Howe, & Stylopoulos (2002)).

H1a: Maximum applied forces will be significantly reduced with the presence of force feedback compared to performance in its absence.

A factorial Analysis of Variance (ANOVA[8]) was conducted with trial number (1-3) and feedback mode (force feedback vs. no force feedback) as independent variables and average maximum forces (N) as dependent variable. The ANOVA found no significant main effect of trial on maximum forces ($F(2,66) = 1.34$, $p = .27$), indicating that the maximum forces that people applied, did not vary significantly over the three trials. The ANOVA did find a medium, yet significant feedback main effect on maximum forces ($F(1,33) = 8.80$, $p<.01$, $\eta_p^2 = .21$)[9], indicating that around 21% of variance in the observed force values is attributable to the addition of force feedback. The mean values indicate that the maximum forces applied to the object were significantly reduced when people received force feedback ($M_{haptic} = 579.86$ N, $SD_{haptic} = 69.71$ N) compared to when they had to rely on the visual display ($M_{visual} = 630.73$ N, $SD_{visual} = 116.70$ N). There was no significant interaction effect ($F(2,66) = 0.87$, $p = .43$), indicating that this force reducing effect of force feedback occurred irrespective of the trial number (see Figure 21). Thus, the data show support for hypothesis H1a.

---

[7] The replacement of outliers with variable means constitutes a common, if controversial method of dealing with outliers, since it can introduce bias into the analysis. Care was therefore taken, that this procedure was only performed if the number of outliers constituted less than 1% of the overall sample and that the identified outliers could be confidently attributed to random causes. Under the condition that these criteria were met, it was judged that the loss of statistical power that would result from the exclusion of cases that contain outliers would increase the probability of committing a Type II error more than the outlier replacement would increase the Type I error probability.

[8] The ANOVA is a parametric statistical procedure that tests the overall fit of a linear model, typically defined in terms of group means, with significance values calculated based on the F-ratio of systematic to unsystematic (error) variance.

[9] η_p^2 is an effect size measure, defined as the proportion of total variability attributable to a particular factor after removing variability due to other investigated factors, plus error variance. Cohen (1988) proposes that $\eta_p^2 = 0.01$ constitutes a small effect, $\eta_p^2 < .09$ a medium effect, and $\eta_p^2 > .25$ a large effect.

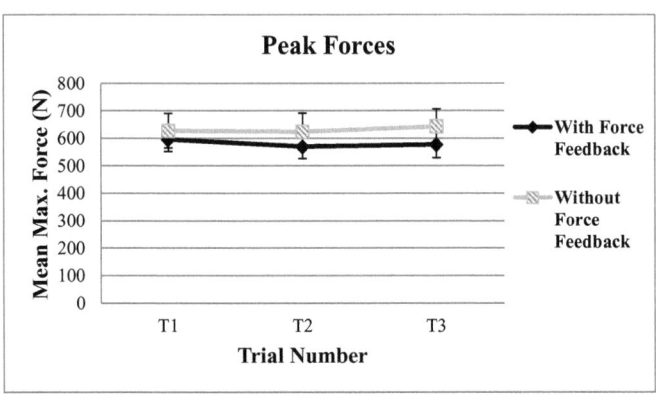

Figure 21. Average maximum forces (N) for trials with haptic force feedback (red) and trials without haptic force feedback (blue).

2.7.2. Regulation of applied forces

H1b: Greater variations in applied forces will occur without the use of force feedback from the remote environment compared to variations in forces applied with the support of force feedback.

Another factorial ANOVA was conducted with trial number (1-3) and feedback type (with haptic force feedback vs. without haptic force feedback) as independent variables and the average standard deviation of applied forces as dependent variables. The ANOVA found a significant trial main effect on force regulation ($F(2,66) = 20.41$, $p<.001$, $\eta_p^2 = .38$). Repeated contrasts of the average standard force deviations of each trial to the previous trial revealed that participants' force regulation decreased in precision significantly in the second trial ($F(1,33) = 39.28$, $p<.001$, $\eta_p^2=.54$), irrespective of the use of force feedback. However, there was no significant difference in average standard force deviations between the second and the third trial ($F(1,33) = 2.54$, $p=.49$). Mean values and standard deviations are listed in Appendix B., Table 8.

The ANOVA further showed no significant feedback main effect ($F(1,33) = 2.38$, $p=.13$) nor a significant interaction effect ($F(2,66) = 1.74$, $p=.18$), suggesting that force feedback had no significant influence on human force regulation as indicated by the average standard deviation of applied forces (see Figure 22). The analysis suggests that hypothesis H1b is not supported by the data.

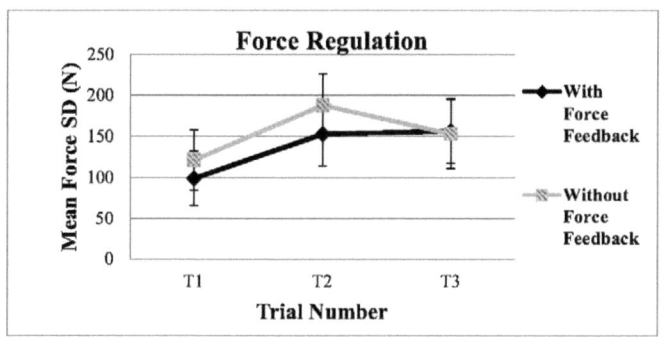

Figure 22. Average force standard deviations (N) for trials with force feedback (red) and trials without force feedback (blue).

2.7.3. Force application precision

H1c: The users' regulation of applied forces will be more accurate to the task requirements when force feedback is present than when it is absent.

Finally, aiming to ascertain whether force feedback had an effect on participants' ability to apply forces of a pre-specified intensity, applied mean forces were investigated. Specifically, it was tested whether mean values were significantly lower than 300 N or significantly higher than 500 N (see Figure 23). Mean force values for each trial and feedback condition and their standard deviations can be found in Appendix B., Table 9.

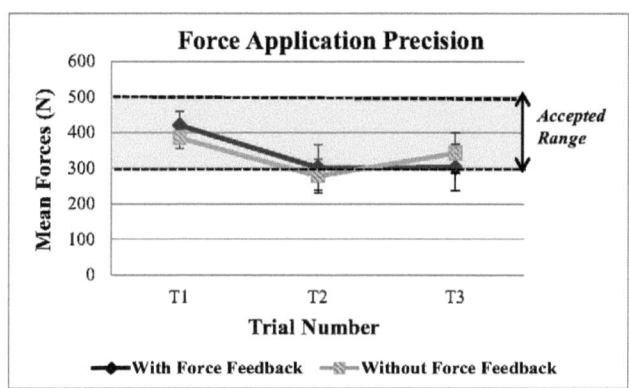

Figure 23. Mean forces (N) for trials with force feedback (red) and trials without force feedback (blue).

Only the mean values for the second trial conducted without force feedback were found to lie outside the accepted range. A one-sample, one-tailed t-test was conducted comparing the mean value of this condition to the test value of 300 N in order to ascertain whether the mean forces were significantly below the accepted range (M = 277.87 N, SD = 96.26 N). The t-test indicated that the mean values did not significantly differ from the test value of 300 N (t(33) = -1.34, p = .38). Hence, it would seem that force feedback did not significantly improve participants' ability to regulate the intensity of applied forces so that they lie within a pre-specified range, as they were equally capable of doing so with a visual display of forces only. Consequently, no empirical support is found for hypothesis H1c.

2.7.4. The relationship between task completion time and measures of force

H2: The effect of force feedback on task performance speed will depend on the performance characteristics of the task (e.g. Massimino & Sheridan (1992); Cao, Zhou, Jones, & Schwaitzberg (2007)). If forces and task completion times correlate significantly, force feedback will increase accuracy, and either increase or decrease completion time, depending on the sign of the correlation. If applied forces and task completion times do not significantly correlate, force feedback will significantly affect force regulation but not task completion time.

In order to determine whether task completion time was influenced by applied forces, bivariate Pearson correlations were calculated between participants' calculated overall task completion times and overall mean forces, maximum forces and the average force standard deviations. These were all found to be non-significant at the $p<.05$ level for trials without force feedback (NF) as well as for trials with force feedback (FF) (see Appendix B., Table 10). It can thus be concluded that people's use of force was not related to the time that they took to complete their task, so that both aspects represent independent indicators of work performance.

A factorial repeated-measures analysis of variance (ANOVA) was then conducted with trial number (1-3) and feedback type (force feedback (FF) vs. no force feedback (NF)) as independent variables and task completion time (tct) as dependent variable. The ANOVA found a significant main effect of trial on task completion time ($F(2,66) = 15.44$, $p<.001$, $\eta_p^2=.32$), regardless of the feedback type. Rather than finding a practice effect, repeated contrasts of the times in each consecutive trial to the previous trial found that participants slowed significantly down for their second trial ($F(1,33) = 29.46$, $p<.001$, $\eta_p^2=.84$), but did not alter their speed significantly for the third trial ($F(1,33) = 0.44$, $p=.51$). Mean values of task completion times and their standard deviations for each trial are listed in Appendix B., Table 11.

Irrespective of this deceleration, the ANOVA found a significant main effect of feedback type on task completion time ($F(1,33) = 10.51$, $p<.01$, $\eta_p^2=.24$). Looking at the mean values, the data suggest that participants completed their task significantly quicker with force feedback ($M_{force\ feedback}$ = 9.94 sec., $SD_{force\ feedback}$ = 5.06 sec.) than without force feedback (M_{visual} = 13.07 sec., SD_{visual} = 6.56 sec.). There was no significant interaction effect between trial and force feedback (F(2,66) = 2.03, p=.14), indicating that participants performed significantly faster with force feedback, regardless of the trial number (see Figure 24).

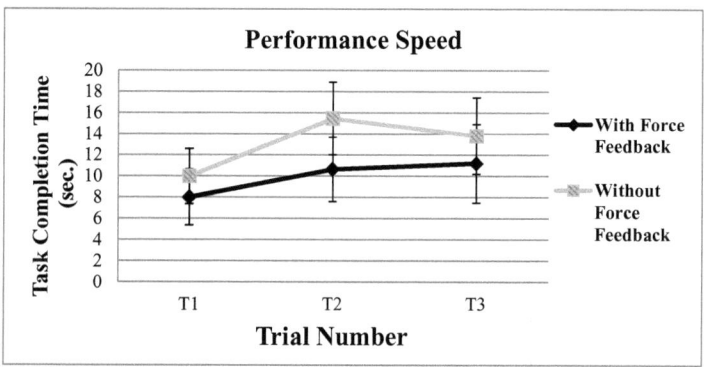

Figure 24. Mean task completion times (sec.) and standard deviations for trials with force feedback (red) and trials without force feedback (blue) in seconds.

To sum up, the results of this experiment indicated no significant relationship between task completion time and force application measures, yet task completion time was significantly reduced with the application of force feedback from the remote environment. It seems that force feedback acts on completion times independently of force reductions; hence, hypothesis H2 was not supported by empirical evidence.

2.7.5. Survey results

Aiming to ascertain user perception of functionality and usefulness, participants were also asked a number of questions. Of 34 participants, all respondents stated that the force feedback portrayed surface characteristics that are encountered in the remote environment realistically. 30 respondents (88.24%) felt that the force feedback was stable and that they felt no friction when they performed the task. 28 participants (82.35%) considered the force feedback a useful aid for this task.

2.8. Discussion

Although the task completion times indicate that participants slowed down their performance after their respective first trials with and without force feedback, the statistical analysis indicated that overall, force feedback allows people to perform their task significantly faster compared to the case where only a visual display of applied forces is available. While participants took on average 13.46 seconds to complete this task without force feedback, they managed to complete their task on average in 9.94 seconds with force feedback, thus reducing task completion time by 26%. Furthermore, even though force feedback enticed the users to accelerate their performance, this did not come at a cost of reduced force control, as speed and force application measures were found to be unrelated for this task. In fact, when receiving force feedback, the users' applied forces were on average more stable, even though this difference failed to reach significance. Moreover, force feedback was clearly found to reduce the peak forces applied to the surface significantly when in contact with the remote environment. Overall, these results corroborate evidence from previous studies, which suggest that haptic feedback can have the effect of reducing the peak forces applied by the human operator in tasks with short-term contact situations (e.g. Wagner, et al. (2002); Mayer et al., (2007)).

Several studies found that performance accelerated with the use of force feedback in cases where task performance speed is positively related to force application accuracy (e.g. Massimino, et al. (1992); Cao, et al. (2007)). On the other hand, no effect on performance was found when there was no obvious link between speed and accuracy (e.g. Deml, Ortmaier, & Seibold (2005); Hurmuzlu, Ephanov, & Stoianovici (1998)). Hence, it was hypothesised that the effect on force reduction would mediate task performance times. However, in the present study, no significant correlation was found between force regulation accuracy and task completion times, yet still a significant acceleration of task performance was identified. This suggests that force feedback from the remote environment may have an independent effect on task completion times, though the underlying mechanism of this effect is, as yet, unclear. Considering that the vast majority of users considered force feedback as an useful aid to this task, it is conceivable that the force feedback simply increased the users' confidence in quickly establishing contact with the steel platform, as this contact was signalled to them with visual as well as haptic cues. Until further investigations on this topic are conducted, the causal factors for faster task performance with haptic cues remain speculative. Furthermore, based on the present experiment, no conclusion can be drawn on whether speed would be adversely affected if it were negatively related to performance accuracy; hence, this predication is subject to future investigations.

Looking at the measurements of applied forces, one may wonder why the absolute intensity of forces would be reduced by the application of force feedback, when force variance is largely unaffected. A possible explanation would be that force and visual displays speak to different task demands. Force regulation requires constant adjustment into two directions, i.e. more or less pressure needs to be applied, as informed by the comparison of the actual value to a norm value. It would seem that visual displays are more apt in conveying relations between two values, i.e. the desired and the actual forces, while force feedback seems more suited to quickly signal the application of excessive pressure. Further studies that investigate this question are needed to validate this assumption.

On the surface, the findings of this study would lead to the conclusion that the use of force feedback devices does offer performance benefits over and above those provided by a simple visualisation of applied forces. While it had no significant influence on the users' ability to apply a constant force within a specified range, force feedback was found to speed up performance whilst at the same time reducing the risk of damage to the material attributable to excessive forces. Specifically, the effect sizes of the statistically significant effects indicate that around 21% and 24% of the variation in peak forces and performance speed, respectively, can be attributed to the use of force feedback. Hence, from the findings of the present study, it may be argued that the use of force feedback can be particularly recommended for cases in which damage to material and slow work performance are major issues of concern, provided that the potential benefits outweigh the additional costs and effort involved in the procurement and implementation of the force feedback device.

3. Guidelines for the implementation and effective use of haptic feedback from the remote environment

Both experiments showed task facilitative effects of haptic feedback interfaces over and above those provided by the best possible visual conditions that are realistically to be expected during comparable tasks and setups. A review of the available literature and the previously discussed experimental results allowed for the derivation of a number of guidelines for the effective use of haptic interfaces in teleoperation systems that should be considered in order to maximise upon their effects.

- Vibrotactile and force feedback signals are particularly recommended in cases where damage to material and equipment due to excessive forces needs to be avoided.
- Force feedback is an effective indicator of force intensity for tasks that require long, continuous periods of contact (app. >80% of task completion time). It is particularly recommended as an addition to visual substitution displays when excessive forces need to be reduced.
- Visual substitution displays are sufficient for informing the user of absolute forces applied and for conveying the notion of force variance.
- When a task requires continuous contact to a surface, force feedback is also advisable for time-critical applications.
- Vibrotactile signals are effective in reducing peak contact forces in tasks that require brief periods of contact (app. < 20% of task completion time), especially under conditions of poor visibility.
- When tasks require contact only intermittently (app. 20-80% of task completion time), vibrotactile feedback does not improve a user's ability to move the manipulator quickly and precisely. Instead, all care should be taken to provide visual feedback with unambiguous cues of relative depth.
- Uniform vibratory signals that provide unambiguous signals seem to be more effective in improving force regulation, error rates and task completion times than directional vibrotactile signals that vary in intensity or location with a change of the stimulus. The latter may even be distracting to individuals and should be avoided.
- Care should be taken that the vibrotactile stimulus is clearly noticeable. It is further advised that a desired action response be conditioned to a vibrotactile stimulus, to ensure that it is correctly interpreted and reacted upon.

Chapter III.
The Effect of Haptic Guidance on Task Performance

While tactile signals from the remote environment may, secondarily to force reduction, also improve the user's movement coordination capability, more immediate ways of achieving this goal have been envisioned. Unlike acoustic and visual displays, haptic in-/output devices can provide their own input for new control commands (either in form of position or force signals), thus actively guiding or passively assisting the user's control of teleoperator (TOP) movements. There are two common uses for this type of haptic guidance. For one, haptic guidance may be used to haptically demonstrate an expert's movements to novices during training in order to transfer learned motor skills and movement strategies to previously untrained users. Second, haptic assistance functions may be used to guide the user's motions with a control during task performance. Empirical evidence to the effectiveness of haptic guidance signals is scarce and ambiguous. Furthermore, since haptic guidance necessarily curtails the user's autonomy over performed movements, the effect of user control on task performance is also of interest, as the literature indicates that it may directly and indirectly affect performance.

Aiming to ascertain whether haptic demonstrations by experts and other haptic assistance functions can effectively improve task performance with teleoperation systems, two further empirical investigations were conducted, in which the effectiveness of haptic assistance functions during task performance and the effects of haptic expert demonstrations on the performance of novice users of teleoperation systems were investigated. In this context, the question of how much control should be taken away from the user to improve task performance and the effect of loss of control on user perceptions of acceptance and usefulness is also addressed.

1. **An investigation into the effect of haptic expert demonstrations on the task performance of novice users with a teleoperated micro-assembly system**

   1.1. **Theoretical approaches to motor skill training**

In many cases, the operation of teleoperation systems may seem less intuitive than the traditional hands-on approach. As a result, effective training of novices on industrial robotic systems becomes increasingly important for employees and employers alike. However, whilst the technology of those systems may be cutting-edge, training methods are often old-school. Currently, training of machinery at the workplace occurs mostly through spoken instructions, observation of experienced users and trial-and-error based learning gained through hands-on practice. As machinery increases in complexity, increasingly calls are made to modernise current training systems (Almansa, Brenner, Wogerer, & Kiriakidis, 2004).

Although increasingly greater effort is made to incorporate theoretical principles and empirical findings on human learning mechanisms into modern training programs (Reznick & MacRae, 2006), it has been noted that theoretically-based research is still mostly lacking in the development of efficient training methods (Salas & Cannon-Bowers, 2001). Training in the use of machinery focuses primarily on motor learning. According to Palmer & Meyer (2000), motor learning refers to mental or physical changes, which are associated with practice or experience, and provide the capability for producing skilled actions. One learning theory on the training of motor skills, which underpins some modern teaching techniques, is Fitts and Posner's (1967) three-stage theory of motor skill acquisition. According to this theory, every new trainee goes through three consecutive stages before he or she finally acquires expertise in a particular motor skill. In the first stage, the cognition stage, the apprentice begins to understand the rules of the task at hand through explanations and demonstrations by more skilled individuals. Performance is erratic at first and occurs in distinct steps. Performance becomes more fluid with fewer interruptions once the student starts to get a grasp on the mechanics involved in the following associative stage. Finally, following deliberate practice and continuous feedback, the behaviour is automated and adaptive, little cognitive input is needed, and the student can focus on refining his or her performance. Towards the end of this final automation stage, the task at hand can be performed with speed and precision.

This theory on motor learning takes into account that human memory can be categorised into two distinct parts: the declarative and the procedural memory (Cohen & Squire, 1980). Declarative memory contains factual knowledge, which can be recalled at will. Procedural memory is largely non-conscious and only accessible through performance of certain behaviours and actions (Wickens

A. , 2000). With regard to learning power, procedural memory is generally considered superior to declarative memory as procedural memories can be acquired, stored, and retrieved without the participation of the limbic/diencephalons brain systems. Thus, this form of memory is "phylogenetically early, reliable and consistent" (Eysenck & Keane, 2000, p. 209). In contrast, declarative memory seems to involve primarily limbic/diencephalic structures combined with the neocortex. This systems is "fast, phylogenetically recent and specialised for one-trial learning" (Eysenck & Keane, 2000, p. 209), but also fallible as it is sensitive to interference and prone to retrieval failures. Thus, although both components of memory are relied upon during machine operation, one might argue that motor skill training should primarily focus on the training of procedural skills.

1.2. Haptic feedback in training simulations

In many disciplines, simulators are used for the training of motor skills (Reznick & MacRae, 2006). High-fidelity simulation systems aim to portray surface textures, equipment failures, motion, vibration and various other situational cues very realistically. Currently, such types of simulators are used primarily in the medical field. In the USA, the Food and Drug Administration panel recommended the use of high-fidelity virtual reality training for carotid artery stenting based on evidence suggesting that these training simulations improved surgeons' performance in these types of surgery (Gallagher & Cates, 2004). Simulators with less physical fidelity were also found to be efficient in the training of key skills (Jentsch & Bowers, 1998). For example, Seymour, Gallagher, Roman, et al. (2002) and Grantcharov, Kristiansen, Bendit, et al. (2004) found that resident physicians who had trained with low-fidelity virtual reality models made fewer intraoperative errors during laparoscopic surgery than did residents who had not benefited from simulation training. While both types of simulators were found to be effective in the training of various key skills, including motor skills, there is a trend to use low-fidelity devices to train complex skills, as these are generally cheaper and easier to operate and maintain (Reznick & MacRae, 2006).

With the advances made in haptic technology over the last decade, various efforts took place to incorporate haptic feedback into simulators and training methods. In fact, there are a number of reasons to suggest that instructing participants with both visual and haptic feedback would be superior to the use of only one modality of instruction. For instance, according to Druyan (1997), several studies found that physical experience creates particularly strong neural pathways. Furthermore, since multi-modal encoding involves more brain structures compared to uni-modal memory encoding, retention of learning might be higher (Wickens A. , 2000). In fact, a number of studies found memory performance after enacting certain actions to surpass memory performance after verbal encoding of the same actions (e.g. Cohen, (1981); Engelkamp & Krumnacker (1980); Engelkamp &

Zimmer (1985)). It is widely assumed that this superiority effect of subject-performed tasks results from multi-modal information processing which is a consequence of performing the actions (Bäckmann & Nilsson, 1991).

Studies on the use of haptic feedback in training abound. There have been a number of reports of cases, in which haptic feedback promoted students' understanding of certain subject matters. For example, Pfister & Laws (1995) and Williams, Chen & Seaton (2002) reported that different haptic devices helped students in grasping basic physics concepts, such as Newton's law. In Greece, a Power Glove with tactile feedback was incorporated into classroom exercises "to build a theoretical model for virtual learning environments, expanding constructivism and combining it with experiential learning" (Mikropoulos & Nikolou, 1996). Okamura, Richard & Cutkosky (2002) used haptic paddles to demonstrate engineering concepts to students. Surveys and opinions of instructors apparently confirmed that students grasped concepts more quickly with this practical demonstration. While these reports can be interpreted as evidence that haptic feedback aids learning of cognitive concepts, without a control group, it is difficult to say whether these examples speak to the value of haptics in particular or simply to that of practical demonstrations in general.

Arguably better-founded claims were made that haptic feedback might be beneficial in the training of general and specific motor skills. In recent years, haptic feedback has been utilised as a learning tool for motor skill training in a wide range of applications, including surgical simulations, molecular docking, and virtual prototyping (e.g. Wagner, Howe, & Stylopolous (2002); Gunn, Hutchins, Stevenson, Adcock, & Youngblood (2005); Ström, Hedman, Kjellin, Wredmark, & Felländer-Tsai (2006)). For example, Seymour, et al. (2002) found that gallbladder dissection was 29% faster for residents who were trained on subtasks that were part of the surgery with a virtual-reality training system with haptic feedback. In contrast, resident surgeons trained the traditional way were five times more likely to injure the gallbladder or singe non-target tissue. Adams, Klowden & Hannaford (2001) reported that participants performed in a real-world Lego assembly task much better when they were trained in a virtual reality environment that included force feedback. However, this effect was not statistically significant. Stredney, Wiet, Yagel, et al. (1998) found that using haptics in a virtual environment increased effectiveness of the training of young rural drivers.

1.3. Haptic expert demonstrations

Since it is generally quite difficult and inefficient to verbalise procedural knowledge, it has been suggested that the demonstration of movements by experts would be an effective way of training users in the handling of machines (e.g. Liu, Cramer, & Reinkensmeyer (2006); Feygin, Keehner, &

Tendick (2002)). Rather than just showing trainees these skills, haptic technology can be employed to make them feel the expert's movements, as well. This can be accomplished by recording the position or force signals sent during execution of a prototypical task with a haptic input/output device, such as a joystick with force feedback capabilities, and playing them back to the trainee (Yang, Bischof, & Boulanger, 2008).

Compelling arguments have been made for the integration of haptic expert demonstrations into training routines. It has been suggested that, in addition to providing information to a second sensory modality and actively involving the trainee, haptic expert demonstrations would also allow the user to learn required muscle activity faster (Liu, Cramer, & Reinkensmeyer, 2006). Hence, haptic expert demonstrations promise to be an effective method for the training of required motor skills. Empirical research evidence on the effectiveness of this training method, however, is equivocal.

In a study by Gillespie, O'Modhrain, Tang, et al. (1997), participants were asked to move and then stop the load of a free-swinging pendulum as quickly as possible, with a shorter stop time considered a better performance. The optimum strategy for this task was demonstrated by a "virtual teacher", a robotic device that moved the participant's hand. The results showed that the learning curves of participants who were taught by the virtual teacher did not significantly differ from those of participants in a control group. The authors surmised that the optimum strategy was perhaps too difficult to master. They proposed, however, that as a "virtual teacher", haptic guidance could encourage subjects to try more advanced strategies of movement. They also suggested that the eye-hand coordination task that they chose overemphasised visual input, and that tasks with a haptic focus might profit more from this type of haptic training strategy.

Haptic guidance also seems to improve timing and force replication. For example, Feygin, Keehner & Tendick (2002) investigated the practical value of haptics for skills training, using the haptic guidance paradigm. Here, participants were physically guided through the ideal motion by the haptic interface. Participants learned complex, three-dimensional motions under three training conditions (haptic, visual, visuo-haptic) and were required to repeat the taught movement. Performance was measured in terms of position, shape, timing, and drift. The results indicated that timing of the movements was more accurately replicated with haptic than with visual feedback, whereas position reproduction was more accurate with visual feedback. In a study by Srimathveeravalli & Thenkurussi (2005), participants were asked to replicate an expert's handwriting by copying shape and force pattern. The results indicated that haptic training promoted force recall but did not promote character-shape learning. Similarly, Morris, Tan, Barbagli, et al. (2007) found that participants recalled forces significantly more accurately following visuo-haptic expert demonstrations than following visual or haptic training alone. The authors propose that combining haptic demonstrations with visual feedback might prove effective in teaching sensorimotor skills with force-sensitive components to them, for instance surgery.

The effectiveness of haptic demonstrations in teaching particular movements is more disputed. Solis, Avizzano & Bergamasco (2003) evaluated the potential of a Japanese character learning system capable of haptic guidance to teach writing skills under three training conditions: visual feedback only, haptic feedback only, and visuo-haptic feedback. Task completion time, overall correction force magnitude, and character recognition probability were used to measure pre- and post training performance. The results indicated that training with haptic feedback only improved task completion time compared to the visual condition, whereas training with visual and haptic feedback showed improvement of all measured indices.

In contrast, Liu, et al. (2006) used haptic guidance in order to teach a group of participants to repeat a certain movement with their hand and arm along a three-dimensional path in space. Since participants were allowed to watch their arm whilst performing this task, visual input was also provided. In the haptic guidance training condition, participants were first guided by a PHANTOM 3.0 haptic device (by SensAble Technologies) along a certain path and then asked to repeat this movement. In the visual training condition, participants simply observed the haptic device as it performed the movement autonomously. Participants' performance in accurately replicating this movement was found to improve with practice, but there was no significant difference in performance between the two training conditions.

In addition to the immediate benefits of haptic expert demonstrations, long-term effects have also been studied. Yang, Bischof & Boulanger (2008) conducted two experimental studies in which they assessed the effectiveness of visual and visuo-haptic demonstrations in developing short-term and long-term motor skills, respectively. Using a PHANTOM Omni device, participants were required to learn three trajectories performed by an expert. In the visual assistance training condition, participants had to trace a trajectory displayed on a screen as they learned the task. With visuo-haptic assistance, participants were also actively guided by the device. Based on their results, the authors concluded that visuo-haptic training offered no significant benefit in terms of short-term motor skill development but argued that the results showed a (non-significant) tendency for benefits in long-term skill development.

By far the largest proportion of empirical studies on haptic demonstrations investigated this method outside of any practical context; instead, most focused on the training of abstract motor skills. On the other hand, current applications of haptic guidance technology to education and industry cover a wide spectrum, from the teaching of calligraphy, in the form of Chinese (Teo, Burdet, & Lim, 2002) or Japanese (Solis, Avizzano, & Bergamasco, 2002) characters, to the training of various surgical skills (Shen, et al. (2008); Williams, Srivastave, Conatser, & Howell (2004)). Yet, most of the evaluations of haptic expert demonstrations in practical applications to date rely on anecdotal evidence and not on empirical performance or learning data acquired through scientifically and methodologically rigorous studies (see Salas & Cannon-Bowers (2001)). Consequently, there is a distinct need for the investigation of the practical value of haptic demonstrations for industrial teleoperation systems under controlled, experimental conditions.

1.4. Research aim

Hence, in order to investigate the practical value of haptic expert demonstrations within the context of an industrial application, another empirical study was designed and conducted. In this case, a teleoperated micro-assembly system was chosen. Specifically, its aim was to investigate whether an expert's demonstration with visual and haptic feedback would be superior in teaching task performance timing and improving task performance compared to a simple video demonstration of the expert's performance. In this context, it was also investigated whether specific cognitive strategies that are crucial to effective task performance would be better impressed upon the trainee with haptic demonstrations. Furthermore, hardly any research had been conducted on whether haptic demonstrations would only impart task-specific motor skills, or whether these acquired skills would generalise to other tasks. Thus, in the event that the haptic expert demonstration improved tracking performance, it was to be investigated whether the use of this haptic demonstration enhances general motor skills or whether the advantage gained by the haptic demonstration only applies to the repetition of the sequence of movements experienced by the haptic demonstrations of the expert's movements with a particular control. Finally, a survey was conducted in order to gauge subjective impressions of the trainees.

1.5. Research hypotheses

The following research hypotheses were investigated:

H1: Users trained with haptic demonstrations show superior motor performance skills to those who were only instructed by video (Solis, Avizzano, & Bergamasco, 2003).

> *H1a: Haptic expert demonstrations are particularly effective in teaching task performance timing (Feygin, Keehner, & Tendick, 2002). It is therefore expected that participants in the visuo-haptic demonstration group will imitate the timing of the expert's movements more closely than those that were instructed by video only.*
>
> *H1b: Haptic expert demonstrations will improve participants' ability to control the tele-operator before they complete their first practise trial.*
>
> *H1c: Since the literature suggests that haptic demonstrations will train specific rather than general motor skills, it is expected that participants with haptic expert demonstrations will perform better than the control group on the demonstrated track, but not on the novel track.*
>
> *H1d: Yet, hands-on practice will have a greater effect on task performance than the type of instruction they receive (Liu, Cramer, & Reinkensmeyer, 2006).*

H2: Haptic expert demonstrations can transfer implicit task knowledge from the expert to the novice (Yang, Bischof, & Boulanger, 2008). The key to successful pick-and-place performance lies in the correct positioning of the gripper before opening/closing the gripper clamps. It is expected that this implicit task knowledge will be conveyed more effectively when the expert's movements are demonstrated with visual and haptic demonstrations, rather than when they just watch the expert perform, thus leading to superior performance.

1.6. Method

1.6.1. Participants

An opportunity sample of 24 male and 8 female participants ($N = 32$) took part in the experiment ($M_{age} = 24$ yrs., $SD_{age} = 3.70$ yrs.), all of whom were right-handed. All but two participants had an engineering background. Participants were assigned to one of two experimental groups (see section 1.6.3). The two groups of participants were completely matched in terms of gender, age, handedness, educational background, ability to coordinate arm-hand movements, and experience in handling joysticks. Post-hoc analyses confirmed that the two experimental groups did not differ based on age ($t_{age}(30) = -.71$, $p = .48$), experience in handling joysticks ($U_{experience} = 112.5$, $p = .55$), or their ability to coordinate arm-hand movements ($t_{coordination}(30) = -.77$, $p = .45$).

1.6.2. Experimental apparatus

The experimental setup is identical to the one described in Ch. II, section 1.6.2 (see Figure 25). The experiment was conducted in collaboration with the Institute for Machine Tools and Industrial Management *(iwb)* of the Technische Universität München, which provided the hard- and software for the experimental setup, as well as implemented the teleoperation system.

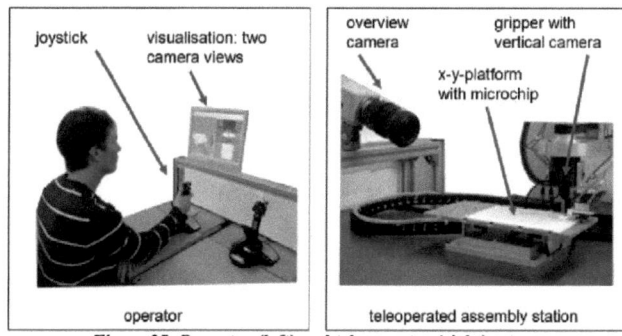

Figure 25. Operator (left) and teleoperator (right) setup.
Source: Reiter, et al. (2008).

1.6.3. Experimental design

In order to test whether operators would learn faster and/or handle the teleoperator more effectively, if they were instructed in the use of the teleoperator via haptic demonstrations of the expert's motions with a control, a 2 (demonstration mode) x 2 (track) x 4 (time) mixed-subjects design was utilised. In one group, participants watched an expert perform a combined tracking/pick-and-place task, whereas participants in the other group were also demonstrated the task haptically, while they watched the expert perform. It was estimated that the expert had over 60 hours of experience in operating this particular system. Care was taken that the chosen task would reflect realistic task demands as they would be encountered outside the laboratory.

For both groups, two types of tracks were used, which the participants were supposed to trace. Only one of the tracks was demonstrated by the expert. The other was similar, but required more advanced handling skills (see Figure 26). The second track was used to assess whether any motor skill learning acquired through the use of haptic demonstrations would be restricted to the repetition of previously demonstrated movement or whether, in fact, acquired skills would generalise to different movements.

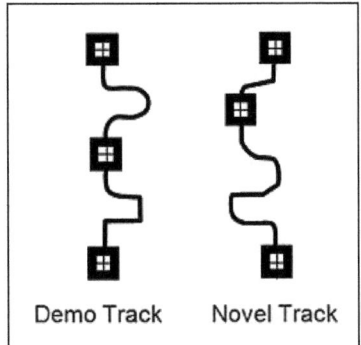

Figure 26. Track demonstrated by expert (left), novel track (right).

The design of the experiment sought to assess effective handling ability with three different measures. Handling efficiency was assessed as task completion time (sec.). As a measure of handling accuracy, position deviation from the track was recorded (mm). Successful grasping of the chip and successful placing of the chip within the confines of the drawn boxes were also assessed as dichotomous measures (successful/unsuccessful). Pick-and-place performance is taken as a measure of handling ability, as the key to successful performance lies in the accurate positioning of the gripper above the planar table: if the gripper's position is too high or too low, the microchip will not be picked up or placed correctly. Beginners tend to have most difficulties with this aspect of the task.

1.6.4. Procedure

Participants were first tested in their hand-arm coordination skills using a tracking test taken from the Motor Skills Test battery of the Wiener Testsystem (by Schuhfried GmbH), and they were asked to estimate their lifetime experience in handling joysticks. Based on their scores in the tracking test and their estimated joystick experience, they were then assigned to one of two groups so as to balance the two groups with regard to joystick experience and arm-hand coordination ability. In one group, participants watched a video showing an expert performing a task in which a microchip is picked up, transported along a curvy line and placed on a specific point along this line. Simultaneously, a video was shown which showed a close-up of the expert's hand as she performed the task (see Figure 27). In the other group, participants watched the same video but also experienced haptic demonstrations, which enabled them to feel the movements performed by the expert. The expert's performance was demonstrated twice to participants. Participants who were only instructed by video were asked to place the palms of their hands flat on the table in front of them to ensure that participants did not mimic the expert's hand movements. Following the expert's demonstration, participants in both groups were asked to repeat the task shown by the expert either on the same track shown in the video or on a previously unknown track. Half of all participants started with the demonstrated track and then performed the task on the novel track; the other participants performed the task with the novel track first and then with the previously demonstrated track. The task was repeated eight times with each track. The design was balanced and the conditions were presented in a systematically randomised order to counterbalance sequence and fatigue effects.

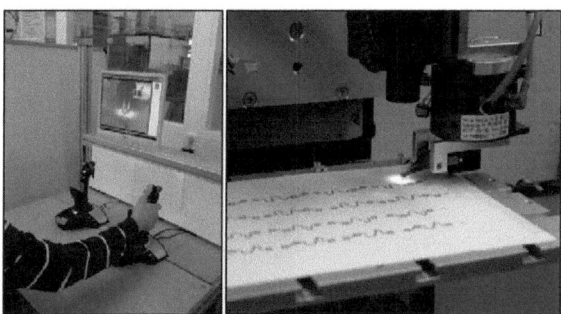

Figure 27. Left: demonstration by expert in the visuo-haptic demonstrations group. Right: teleoperator and test tracks.

1.7. Results

Data were inspected for outliers and the assumptions for parametric tests were tested (see Appendix A.). Where the assumptions were met, data with z-scores of greater than 3.29 were considered outliers and consequently excluded from further analysis. In cases in which the assumptions were violated, outliers were identified via boxplots and corrections or non-parametric tests were applied.

1.7.1. Timing of movements

H1: Users trained with haptic demonstrations show superior motor performance skills to those who were only instructed by video (Solis, Avizzano, & Bergamasco, 2003).

> *H1a: Haptic expert demonstrations are particularly effective in teaching task performance timing (Feygin, Keehner, & Tendick, 2002). It is therefore expected that participants in the visuo-haptic demonstrations group will imitate the timing of the expert's movements more closely than those that were instructed by video only.*

First, it was investigated whether those who were instructed with visual and haptic demonstrations by the expert would copy the timing of the expert more accurately than those who only watched the video. Since half of all trials were conducted with a novel track, only those trials were assessed which the expert had previously demonstrated to the participant. Mauchley's test of sphericity[10] indicated that the assumption of sphericity had been violated for the practice main effect ($\chi^2(27)$ = 145.73, p<.001). Consequently, degrees of freedom were corrected using Greenhouse-Geisser[11] estimates of sphericity (ϵ = .45).

A mixed-subject ANOVA with instruction group as between-subject variable and number of trial as repeated-measures variable found a significant practice effect on time deviation ($F(3.17, 95.11)$ = 12.72, p<.001, η_p^2 = .30). Mean absolute time deviations and their standard deviations for each group are depicted in Figure 28.

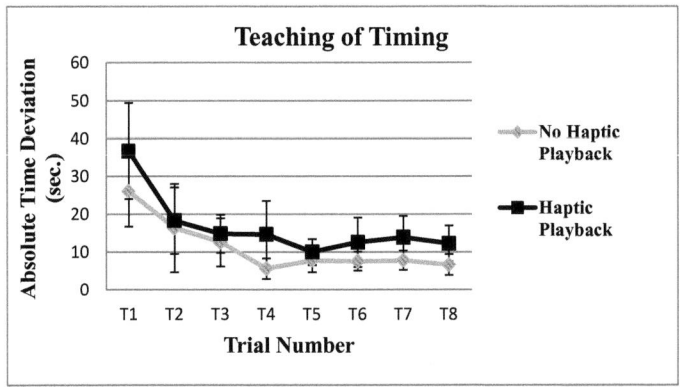

Figure 28. Mean time deviations from the expert's demonstration.

There was no significant main effect of instruction type on participants' timing ($F(1,30)$ = 3.86, p = .06), nor a significant interaction effect ($F(7,210)$ = 0.58, p = .78). Overall, the data suggest that those instructed haptically did not copy the expert's timing more accurately than would those that were only instructed by video. Hence, the stipulated hypothesis (H1a) is not supported. Instead, it appears as if practice resulted in task completion times that increasingly approximated that achieved by the expert, irrespective of the type of training received.

---

[10]Mauchley's test compares the variance-covariance matrix of the observed data to an identity matrix. If the variance-covariance matrix is not a scalar multiple of the identity matrix, the parametric assumption of sphericity is not met and a correction must be applied to the degrees of freedom of the F-ratio in an ANOVA (Field, 2009).

[11] The Greenhouse-Geisser correction adjusts the degrees of freedom used to assess the observed F-ratio. The closer ϵ is to 1, the closer the data are to being spherical (Greenhouse & Geisser, 1959). The correction tends to lead to conservative estimates of the F-ratio if ϵ > 0.70 (Huynh & Feldt, 1976).

1.7.2. Task performance improvement

H1b: Haptic expert demonstrations will improve participants' ability to control the teleoperator before they complete their first practise trial.

H1c: Since the literature suggests that haptic demonstrations will train specific rather than general motor skills, it is expected that participants with haptic expert demonstrations will perform better than the control group on the demonstrated track, but not on the novel track.

H1d: Yet, hands-on practice will have a greater effect on task performance than the type of instruction they receive (Liu, Cramer, & Reinkensmeyer, 2006).

Position Inaccuracy

While the lack of data on participants' deviation from the provided path prohibits any investigations regarding possible beneficial effects of haptic demonstrations on movement precision, it can be investigated whether haptic demonstrations reduce position inaccuracy by comparing the participants' task completion times in each group to those of the expert. Under the consideration of Fitts' Law which stipulates a speed/accuracy trade-off in aimed movements, this investigation is based on the assumption that the expert achieves the fastest possible task completion time at the highest degree of precision. If this assumption holds true, novice users would necessarily be less accurate in their movements, if they performed the task in less time than the expert. Note, however, that no inference regarding position inaccuracy may be drawn from longer task completion times, as a novice may be equally or still less accurate than the expert when they take longer to complete the task. It is also important to note that this time differential only provides an indication of inaccuracy, not a measure of accuracy. That is, ANY task completion time that is faster than that of the expert indicates inaccuracy, whilst no information can be derived regarding the EXTENT to which novice users were inaccurate. As such, a direct comparison in time differences between the two groups would not be informative. However, by looking at the negative difference between performance times of novice users in each group and comparing it to that of the expert, it may be tested, whether participants in both groups performed significantly more inaccurate than the expert, or whether only participants of one group (if at all) would be significantly more inaccurate than the expert. Figure 29 displays means and standard deviations of negative differences in task completions times to that of the expert.

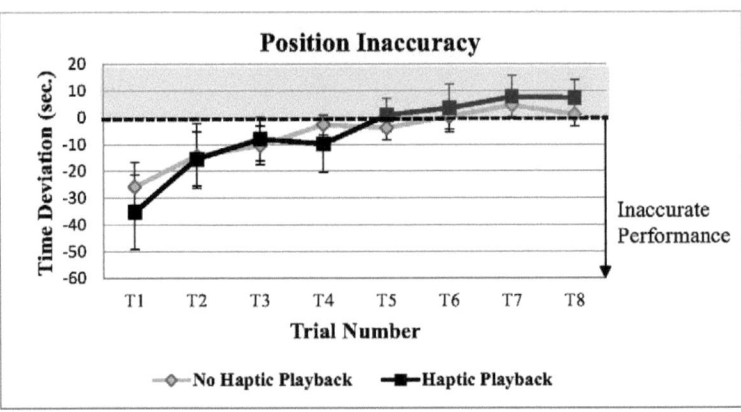

Figure 29. Position inaccuracy in task performance for both demonstration groups as indicated by the negative difference in task completion times of each participant to that of the expert.

Looking only at deviations in task completion time of less than 0 sec., Bonferroni-adjusted one sample t-tests comparing the mean negative deviations of each trial to a test value of 0 found that, for both groups, the task completion times for the first three trials were significantly faster and consequently must be more inaccurate than that of the expert (at p<.006). The results of the individual t-tests for both training groups can be found in Appendix C., Table 12.

Thus, the results would suggest that for the first three trials, users were significantly more inaccurate than the expert was, regardless of the type of playback they received. Hence, this analysis shows no support for hypothesis H1b. At the same time, it offers tentative support for hypothesis H1d in that practice, but not instruction type, reduces position inaccuracy.

Task completion time

In order to ascertain the effect of haptic expert demonstrations on task performance improvement, a mixed-subjects ANOVA was conducted for task completion times, with training group (haptic demonstrations/no haptic demonstrations) as between-subject variable, track type (demonstrated/novel track) as within-subject variable and trial number (T1-T8) as repeated-measures variable. Mauchley's test of sphericity indicated that the assumption of sphericity had been violated for the practice main effect ($\chi^2(27) = 84.48$, $p<.001$) as well as the practice*task interaction effect ($\chi^2(27) = 107.98$, $p<.001$). Consequently, degrees of freedom were corrected using Greenhouse-Geisser estimates of sphericity ($\varepsilon_{practice} = .55$, $\varepsilon_{practice*interaction} = .42$).

A group main effect for task completion times was non-significant ($F(1,30) = 0.15$, $p=.70$). Hence, the statistical analysis showed that participants who witnessed the expert's performance visually and haptically did not perform faster than those who only watch the expert perform. This analysis does not support hypothesis H1.

In order to test hypothesis H1b with regard to task completion times, it was investigated whether training with haptic demonstrations of expert movements would improve participants' ability to control the teleoperator before they completed their first practise trial. For this purpose, the mean task completion times for the first practical trials on the demonstrated and the new track were investigated. If haptic training improved the users' ability to control the teleoperator significantly, the task completion times of participants' first trials should differ significantly, depending on the type of training that they had received. Since for each training group, half of all participants started with the demonstrated track and half with the novel track, it was first investigated, whether the times for the respective first trials differed significantly, depending on the type of track used. This was not found to be the case ($t(30) = 1.07$, $p = .29$). Hence, the times of the first trials were combined so that for each group, 16 cases could be investigated rather than just eight. The mean task completion times for the first trials indicated that, on average, participants in the haptic demonstrations group took slightly longer to perform their task ($M_{haptic} = 109.75$ sec., $SD_{haptic} = 24.32$ sec.) compared to those who did not receive haptic training ($M_{visual} = 97.81$ sec., $SD_{visual} = 17.49$ sec.). An independent t-test indicated this difference to be non-significant ($t(30) = -1.59$, $p=.12$). This finding would suggest that haptic training did not give participants a significant advantage with regard to performance speed, thus hypothesis H1b is not supported by the data.

A three-way interaction effect between practice, track type and training group was also non-significant ($F(7,210) = 0.80$, $p=.59$). Considering the previously reported non-significant group effect and the significant practice effect, this analysis suggests that the previously reported lack of effect of the training group was found for both track types: the one that was demonstrated by the expert and the track that was previously unknown to the participants. Instead, users improved with practice, regardless of the type of training that they received, or whether the track was new or previously demonstrated to them. Hence, the results show no support for hypothesis H1c. Mean times for each trial for the demonstrated and the novel track are depicted in Figure 30 and Figure 31, respectively.

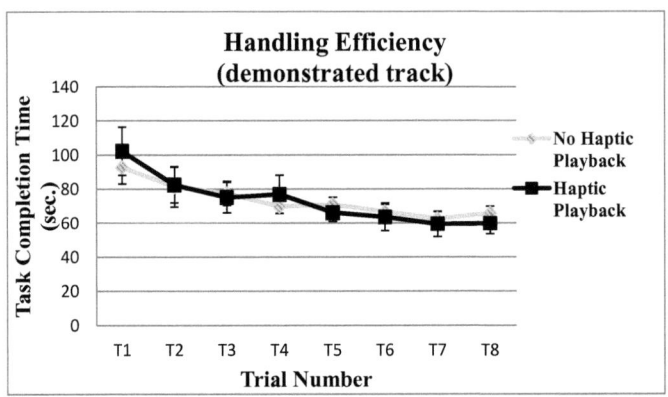

Figure 30. Mean task completion times with standard deviations for the track previously demonstrated by the expert.

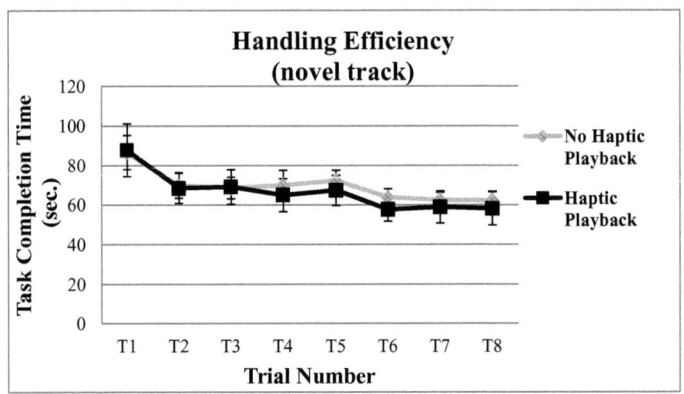

Figure 31. Mean task completion times with standard deviations for the novel track.

The ANOVA further indicated a significant main effect of practice ($F(3.84, 115.13) = 561.74$, $p<.001$, $\eta_p^2 = .65$) on task completion time, but not a significant interaction effect of practice and group ($F(7,210) = 1.67$, $p=.12$). Hence, the analysis offers support for hypothesis H1d.

1.7.3. Implicit task knowledge

H2: Haptic expert demonstrations can transfer implicit task knowledge from the expert to the novice (Yang, Bischof, & Boulanger, 2008). The key to successful pick-and-place performance lies in the correct positioning of the gripper before opening/closing the gripper clamps. Hence, it is expected that this implicit task knowledge will be conveyed more effectively when the expert's movements are demonstrated with visual and haptic demonstrations, rather than when they just watch the expert perform, thus leading to superior performance.

For this purpose, the number of successful and unsuccessful pick-and-place attempts were recorded for each person and evaluated by a trained rater. A picking attempt was considered successful if the microchip was grasped securely enough by the gripper clamps, so that it could be transported the entire track without dropping it. A placing attempt was considered successful, if the microchip was placed entirely within the black lines of the target box. Whenever the camera's view was blocked or there was any doubt regarding the participant's success, attempts were not counted. Subsequently, picking and placing attempts for the demonstrated and the novel track were converted into percentages. Mean and median percentages and the standard deviations are shown in Appendix C., Table 13. The values indicate that, on average, participants who were instructed with visuo-haptic expert playback performed slightly worse than those who only watched the expert.

Mann-Whitney tests comparing pick-and-place performance between the two playback groups showed that pick-and-place performance did not significantly differ between these two groups for the demonstrated track ($U_{picking}$ = 95.50, p = .23; $U_{placing}$ = 82.00, p = .09). For the novel track, haptic demonstrations were also not found to have a significant effect on picking or placing performance ($U_{picking}$ = 112.50, p = .58; $U_{placing}$ = 115.50, p = .65).

Again, assuming that practice might overshadow the performance enhancing effects of the expert's demonstration, it was also investigated, whether the type of instruction that participants had received would affect their performance on their very first picking or placing trials. Fisher's exact test found no significant association between the type of training someone received and whether or not the chip would be successfully picked up ($\chi^2(1)$ = 0.82, p =.65). Similarly, no significant association was found between training type and chip placing performance ($\chi^2(1)$ = 1.78, p =.25). Overall, the analyses show no support for hypothesis H3.

1.7.4. Survey results

Finally, towards the end of the experiment, all of the participants were administered a questionnaire in which they were asked under conditions of anonymity, whether they believed the expert's demonstration of the task helped them in performing their task. Of those who only watched the video, seven out of 16 participants believed that the expert's demonstration helped them in performing their task, compared to only two out of 16 participants who, in addition to watching the video, also experienced the expert's movements haptically. In contrast, four participants stated that the video-only demonstration of the expert did not help them, whereas more than half (N = 9) of the participants in the haptic feedback group decided that the demonstration did not help them. In each group, five participants were undecided in this matter (see Figure 32).

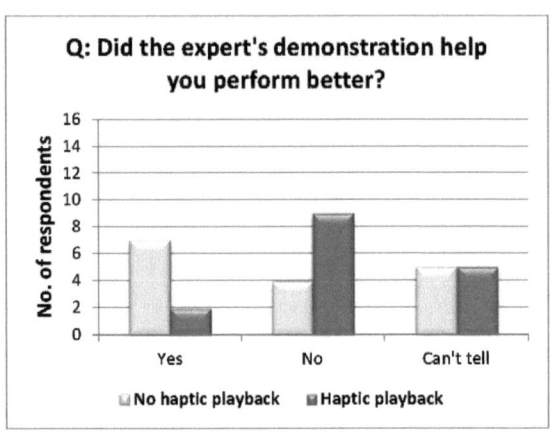

Figure 32. Number of people (out of 16) stating whether they believed the expert's demonstration helped them.

In an open follow-up question, the most frequent comments that participants had made were that they thought hands-on practice helped them more and that they would have preferred to start with practice trials rather than the expert's demonstration.

1.8. Discussion

A number of studies have investigated the effectiveness of haptic expert demonstrations, in which trainees experience an expert's movements haptically. In the literature, it has been suggested that haptic expert demonstrations may not only be beneficial for motor training, but that it may also be superior to visual observations of experts in teaching movement timing and cognitive strategies. While a wide variety of practical applications are proposed, most studies to date focused on teaching abstract movements. The purpose of the present study was therefore to evaluate the use of haptic expert demonstrations within the context of a practical application, i.e. the operation of a teleoperated micro-assembly system. It was also investigated whether these three aspects of motor skills, timing and strategical manoeuvring, would only be improved with regard to the specific task demonstrated by the expert or whether improvements would generalise to another, yet conceptually similar, task. Finally, trainees' subjective impressions of the expert's demonstrations were gauged.

The statistical analysis of the results showed that none of the five main hypotheses were confirmed. It must be noted that due to the fairly small sample size, the (a posteriori) test power, i.e. the probability of detecting a present effect, was in many cases not sufficient to detect small effects. However, since the focus of the presented studies lies on the present-day practicality of the investigated applications, one might argue that only medium to large effects are likely to improve performance to the extent to which it will have noticeable economic effects. Hence, although technically, non-significant effects cannot be considered as evidence that a particular application is ineffective, it may be interpreted as evidence that the investigated application is not likely to lead to practically relevant performance improvements.

Judging from a comparison of participants' task completion times to that of the expert, it seems that experiencing an expert's movement haptically while watching her perform did not lead participants to imitate expert's timing of movements any more closely than did the visual observation itself. This finding was not expected as some studies, most notably Feygin, et al. (2002), observed that haptic demonstrations lend themselves particularly well to teaching temporal aspects of dynamic movements. However, Feygin, et al. also noted that the presence of visual feedback increased position accuracy and that when the task emphasised vision, visual input would dominate and "override" the effects of haptic training. It is debatable whether the task utilised in the present study would force users to focus on correct positioning, thus neglecting the aspect of timing. In fact, this seems unlikely given that trainees in both demonstration groups were much faster than the expert was in the beginning and approximated the timing achieved by the expert as they became more practised. It should also be noted at this point, that participants were told that their timing as well as their tracking accuracy would be assessed. Assuming that they were less skilled than the expert was, this finding would suggest that participants improved in timing, as well as position accuracy with increasing practice. Thus, it would seem that while it cannot be ruled out that haptic demonstrations would teach movement timing effectively in work environments that offer no or inadequate visual feedback, the present study suggests that when visual feedback is present and precise timing is required, haptic demonstrations do not eliminate the need for sufficient hands-on practice.

Practice was also found to be more effective than haptic demonstrations when it came to motor skills and cognitive strategies. These aspects were assessed by tracking performance and picking/placing success. Looking at tracking performance, trainees who were instructed with visual and haptic demonstrations of the expert's movements profited from practice to the same degree as those who only watched the video, as they produced similar task completion times and movement inaccuracies. Conversely, before participants had any practice in performing the task themselves, the haptic demonstration did not give trainees an advantage over those who lacked this experience. This finding applied to the previously demonstrated task as well as a novel task, which had the same task requirements but required different movements. Hence, the data indicate that haptic demonstrations do not significantly accelerate the process of teaching a person to handle a teleoperated device effectively. This is supported by the questionnaire data, in which over a third of participants indicated that the demonstration by the expert did not help them in performing their task and it was suggested by several that they would prefer hands-on practice. However, it should be kept in mind that participants were only demonstrated the movements twice. If participants had witnessed more haptic demonstrations, it cannot be ruled out that the haptic demonstrations of the expert's movements might eventually achieve an effect similar to that achieved by practice, as it was suggested by Solis, et al. (2003). In their study, however, participants' performance after haptic expert demonstrations was compared to pre-training performance scores. Thus, the effect of haptic expert demonstrations cannot be differentiated from that of practice with the device. In order to make that distinction, a control condition would be required, in which participants practiced without haptic demonstrations of the expert's movements.

It could be argued that, with a lack of data on teleoperator positioning, task performance can only be judged with regard to speed, not accuracy, and hence no conclusions can be made regarding the effectiveness of haptic movement demonstrations in teaching movement accuracy. However, if the premise is accepted that the expert would perform the task with the greatest possible speed and accuracy, one may infer that beginners cannot achieve a similar level of accuracy in a shorter completion time. Consequently, task completion time can be considered a measure of inaccuracy insofar that shorter task completion times necessarily indicate movement imprecision. The findings suggest that, regardless of whether or not they experienced the expert's movements haptically, beginners were significantly more inaccurate than the expert for at least the first three trials. Whether or not movement precision increased beyond those three trials for any or both groups cannot be said. Yet, it would stand to reason that if haptic demonstrations improved movement precision, this would become apparent in the first trials, not the later ones. Thus, the present study offers tentative evidence that haptic expert demonstrations are not more effective in increasing trainee's precision of movements than visual observation of an expert's demonstration. This finding also concurs with Liu, et al. (2006) and Yang, et al. (2008), the latter of whom concluded in their study that visuo-haptic training is not as effective in teaching movements as once thought.

A similar conclusion would seem to apply to the teaching of cognitive strategies, where expert demonstrations promise to hold an advantage to practice. It was argued that experts would have developed certain strategies during their many hours of practice that would optimise performance. In observing these strategies in the expert's behaviour, the oftentimes laborious process of developing these strategies could be spared to the beginner (e.g. Gillespie, et al., 1998). The present study investigated whether these strategies are acquired more effectively through the use of haptic demonstrations compared to visual observation of the expert. In the present experiment, the correct positioning of the gripper clamps during picking and placing of the microchip had been identified as a cognitive strategy that the expert had developed during practice. A comparison of participants' pick-and-place performance showed no significant difference in performance of users that were demonstrated the task visually and haptically and those that only watched the expert perform. It would seem that this cognitive strategy was not recognised as such by participants in both groups, or that visual observation suffices to learn it from the expert. Further study seems warranted to investigate the use of haptic demonstrations in teaching cognitive strategies.

Many researchers that found haptic demonstrations to be ineffective remarked that the task that they chose was to blame. Yokokohji, et al. (1996) thought perhaps their chosen task could be learned too quickly and therefore did not require expert knowledge and skills to master. Others (e.g. Liu, et al., 2006) believed some tasks to be too complex for haptic demonstrations to show any discernible effect. Since it was the aim of this study to investigate the value of haptic expert demonstrations in an industrial context, the task used in the present study was designed with realistic task requirements, as they would be encountered at work. Thus, it emphasised accurate timing as well as accurate positioning, with visual feedback present. For such tasks, haptic demonstrations do not seem to offer any discernible advantages over simple videos of demonstrations by experts.

To sum up, haptic expert demonstrations might transfer cognitive strategies from the expert to the trainee; however, the evidence remains inconclusive as to whether haptic demonstrations can accelerate the learning process that is required in order to maneuver a teleoperator arm effectively. Looking at Posner and Fitts' (1967) three-stage theory of motor learning, expert demonstrations only seem to speak to requirements of stage one: the cognitive stage. Apparently, advanced stages of associative learning and automaticity require different training practices. Yang, et al. (2008) also suggested that haptic expert demonstrations might not be effective in improving motor task performance, as it encourages trainees to be passive. Since passive learning has been widely found to be less effective than active learning (e.g. Srimathveeravalli & Thenkurussi, 2005), this explanation seems plausible. It seems therefore plausible that that overall, performance skills would be better acquired through hands-on practice. Hence, the results of this study suggest that taking too much control away from the user during motor skill training may be counterproductive to the learning of movement skills.

It has also been noted that haptic expert demonstrations lack feedback on the quality of performance and individual errors, another important element of learning (Eysenck & Keane, 2000). As Yang, et al. (2008) have suggested, it may be sensible to promote learning with haptic training strategies that force the participant to be active, coupled with demonstrations of and feedback on positive and negative displays of performance. Whether the development and implementation of such training strategies are worth the effort and investment remains to be assessed.

2. An investigation into the effect of haptic assistance functions on task performance and user perception with a virtual teleoperation system

Another means, by which haptic feedback promises to enhance movement coordination with teleoperation systems, comes in the form of haptic assistance functions. These may be employed to facilitate accelerating or decelerating movements, actively pull the user along a given path, prevent them from deviating from a path and/or prevent them from entering „forbidden regions" (Rosenberg L. B., 1993). A wide range of real-world applications were found suitable for the use of haptic assistance. For instance, haptic assistance functions have been used to limit the motion of a tool to a desired movement range for applications in surgery (e.g. Park, Howee, & Torchiana (2001); Marayong & Okamura (2004)), as a technique to control dynamic tasks such as driving (Forsyth & MacLean, 2006), and as a means of increasing movement precision in teleoperated assembly lines (e.g. Nakamura & Honda (2006); Takesue, Murayama, Fujiwara, Matsumoto, Konosu, & Fujimoto (2006)). A comprehensive overview of assistance functions applied to teleoperation systems and physical human-robot interaction tasks is given in Passenberg, Peer & Buss (2010).

In theory, haptic assistance functions aim to capitalise on the accuracy of the robotic system, while still allowing for human flexibility in the performance of tasks (Abbott, Hager, & Okamura, 2003). Different assistance functions can provide differing levels of guidance to the user, ranging from complete guidance to no guidance at all. As such, assistance functions, although conceptually similar, come in various guises. Virtual fixtures (e.g. Aarno, Ekvall, & Kragic (2005); Lin, et al., (2006)), virtual mechanisms (e.g. Joly & Andriot (1995); Micaelli, Bidard, & Andriot (1998)), virtual tools (Itoh, Kosuge, & Fukuda, 1995), virtual paths and surfaces (Moore, Peshkin, & Colgate, 2003), and haptically augmented teleoperation methods (Turro, Khatib, & Coste-Maniere, 2001) have been applied to teleoperation systems using a variety of methods.

The various forms of haptic assistance can be subsumed under two functional categories: i.e., those that provide task-dependent assistance and those that can offer assistance independently of task knowledge. For instance, virtual fixtures use software-generated force and position signals to increase movement accuracy of the teleoperator, whereby assistance is provided based on the (tele-) operator's absolute position in space in reference to its desired position or obstacles in the remote environment. Consequently, these types of assistance functions are generally task-dependent since desired paths or potential fields need to be pre-defined for the environment (Abbott, Marayong, & Okamura, 2007). In the case of virtual fixtures, a further distinction may be made between active and passive fixtures. Passive virtual fixtures scale the force applied by the user, for instance in order to drive the user back to a desired path (Abbott, Marayong, & Okamura, 2007). Contrary to passive fixtures, active fixtures apply forces into the desired direction, even if the operator applies no force at all (Pezzementi, Okamura, & Hager, 2007).

Task-independent assistance functions, e.g. variable admittance/impedance controllers, may assist (tele-)operator control irrespectively of task objectives, for instance, by facilitating acceleration and deceleration manoeuvres based on the prediction of users' intended movements (Duchaine & Gosselin, 2007). In this case, explicit knowledge of the task objectives not required. Thus, task-independent assistance functions tend to offer the user more autonomy over movements compared to task-dependent assistance; however, in return they usually provide less support in performing a task quickly and accurately.

2.1. Previous studies on the effects of haptic assistance on task performance

Numerous studies have been conducted on task performance-improving aspects of haptic assistance functions. For instance, Duchaine and Gosselin (2007) proposed a task-independent form of haptic assistance using a variable impedance control where the human's intention to move is predicted based on the time derivative of the force applied in order to move the operator. They tested in various tasks whether the assistance function would facilitate users' movements with a cooperative robot. Although the experiments lacked sufficient sample size for a plausible generalisation of their results beyond the theoretical confines of these studies, they indicate that the proposed form of task-independent haptic assistance may have the potential to improve cooperative movement coordination with the robot in pick-and-place and maze navigation tasks.

The majority of studies on performance-related aspects of haptic assistance focused on the use of task-dependent assistance functions, in particular virtual fixtures. Rosenberg (1993) implemented haptic virtual fixtures, consisting of virtual walls with stiffness and damping properties for a tele-operated peg-in-hole task. As a measure of task performance, Rosenberg calculated a performance index comprised of task completion times and task difficulty. User experiments indicated task performance improvements of as much as 70% with the use of the fixtures. Teo, Burdet, & Lim (2002) virtually attached a damped spring to the tip of their interface in order to teach Chinese handwriting. They observed increased movement and force accuracy for each of their three subjects, and noted that the fixture was „agreeable" to users. Steele and Gillespie (2001) studied haptic guidance in driving and its effect on performance and visual demand. They showed that haptic feedback reduced the risk of deviation from the given path by 50% and decreased visual demand on the driver by 42%. Payandeh and Stanisic (2002) implemented virtual fixtures to guide the users and to prevent them from entering forbidden areas. They found that the virtual fixtures reduced the time it took to complete a teleoperated assembly task and improved the users' precision. Moreover, the users' workload seemed to have been reduced, as was the training time for novice operators. Similarly, Ren, Patel, McIsaac, Guiraudon & Peters (2008) tested haptic guidance with five people in a task that required the rapid removal of tissue using a rod-like surgical tool. They observed that users required considerably less time (a reduction of 55%) when performing this task with haptic assistance.

Marayong & Okamura (2004) used haptic path guidance to provide assistance in human-machine cooperative systems particularly suited for small-scale tasks. They, too, reported faster completion times and improved accuracy with the use of haptic assistance functions. Moreover, the authors also studied the effect of the level of guidance on task performance in a tracking task. As it would have been expected, task completion times were found to decrease with increasing guidance. Furthermore, when the objective was to follow the path, deviations tended to decrease with increasing guidance; however, when the task objective required to steer away from the path in order to reach an off-path target, deviations tended to increase with increasing guidance, even though the latter effect was non-significant.

2.2. Haptic assistance vs. user autonomy

The experiment by Marayong, et al. (2004) demonstrates fittingly the difficulty in evaluating haptic assistance functions and their effects on task performance. That is, previous studies on haptic assistance functions tended to focus on technical properties of virtual environments and haptic devices, and, to a smaller degree on their effects on the user's motor task performance. Yet, one might argue, that measuring task performance accuracy and speed by itself is not sufficient in an evaluation of the concept. After all, based on the measures of accuracy and speed, the best task performance would be most likely achieved with fully automated systems. However, teleoperated systems would be employed in situations in which automation is not feasible, the flexibility and reactions of humans are desired, or when there is a possibility that the virtual fixture may be computed incorrectly based on faulty sensor data. In these cases, a strong assistance, that takes movement control away from the user, might be counterproductive with respect to work performance. A central question for the development and employment of haptic assistance functions is, therefore: how much assistance should be provided during teleoperation?

Certainly, from a technical standpoint, the extent to which haptic assistance functions can assist the user's movements is stipulated by the demands of the system and the environment. To date, strong assistance can only be provided, if the task demands and features of the environment are well known, and if the system's force output capability can match or surpass that of the human user. Yet, even if these premises are met, it is not certain that an increase in assistance would lead to a corresponding improvement of the user's movement coordination with the teleoperator in the remote environment.

Moreover, it seems that movement autonomy might not only affect the user's performance, but also the user's level of comfort in working with the haptic assistance functions. Compared with visual and auditory signals, haptic signals are generally more intrusive as they may directly interfere with the users' intended movements, by actively inhibiting or restraining them. A number of studies in the areas of social, clinical, and engineering psychology suggest that the experience of a loss of autonomy or a restriction of motor control is perceived as uncomfortable and may even lead to frustration and distress (e.g. Lefcourt (1973); Scepkowski & Cronin-Golomb (2003); Syrdal, Koay, Walters, & Dautenhahn (2007)). Apart from the fact that it is obviously undesirable to subject employees to discomfort and stress, comfort in working with safety-enhancing applications is also an important indicator of their acceptance, and consequently use during the work process (Glendon, Clarke, & McKenna, 2006).

Bias in perception may also affect the way in which work with a particular system is evaluated by its user(s), as studies on human-robot interaction suggest that the level of control that the user is given, changes the perception of performance with this system. For instance, Kim & Hinds (2006) found that users tend to blame and credit the robot for faulty and successful behaviour, respectively, if this robot displays a high degree of autonomy. On the other hand, users seem to credit and blame themselves, if they believe that they were in control of the robot's behaviour. Furthermore, the social psychology literature offers sound empirical evidence for a particular attribution bias, whereby people tend to attribute behaviour of others to individual factors, whilst attributing their own behaviour to situational factors (Hogg & Vaughan, 2002). This attribution bias may result in different perceptions of a system's behaviour, depending on whether this behaviour is attributed to oneself or another. Thus, it seems likely that the perception of working comfort as well as task performance would be influenced by the level of movement control that an assistance function affords the user over TOP movement. For the design of future assistance functions in general, and haptic assistance functions in particular, it is hence of particular interest to investigate both performance-, as well as perception-related aspects of this assistance during teleoperation.

2.3. Research aim

An experimental user study was conducted that aimed at evaluating the effect of control autonomy that is provided by haptic assistance functions on task performance as well as user perception. The chosen task was designed to reflect realistic demands of teleoperated delivery, such as it is currently practiced in some warehouses around the US, where robots are used to quickly transport merchandise around various obstacles (KIVA Systems, 2011). In this context, three different assistance functions, adopted and slightly modified from literature, have been investigated. The three assistance functions were designed to provide different degrees of autonomy over teleoperator movements. Furthermore, two of these assistance functions offered task-dependent assistance, whilst one provided task-independent assistance, allowing for a comparison of effectiveness and perception between these two different types of haptic assistance.

2.4. Research hypotheses

The following research hypotheses have been tested:

H1: Task-dependent assistance functions are necessarily more imposing than task-independent assistance functions (Abbott, Marayong, & Okamura, 2007). Hence, the four conditions of assistance will differ in users' perceptions of their control over TOP movements, with task-dependent assistance functions being perceived as taking more control away from the user than task-independent assistance functions.

H2: Provided that the task and the environment can be precisely pre-defined, task-dependent assistance functions can provide greater assistance than task-independent assistance functions (Abbott, Marayong, & Okamura, 2007). Since this condition applies in the present experiment, assistance functions will lead to greater improvement in objective measures of task performance, the less control is provided to the user.

H3: People feel uncomfortable when their control over their movements is curtailed (Lefcourt (1973); Scepkowski & Cronin-Golomb (2003); Syrdal, Koay, Walters, & Dautenhahn (2007)). Hence, it is predicted that the level of comfort will be determined by the perceived amount of control that the assistance functions afford the user. Specifically, it is expected that users will feel less comfortable working with an assistance function that curtails their movement autonomy strongly, and feel more comfortable when performing with more movement control.

H4: The perceived level of control provided over TOP movements can influence the way people perceive their performance with an assistance function (Kim & Hinds (2006); Hogg & Vaughan

(2002)). It is therefore predicted that performance judgements will deteriorate with decreasing levels of control provided to the user.

2.5. Method

2.5.1. Participants

An opportunity sample of 16 male and 16 female (N=32) participants (M_{age}= 24.61, SD_{age}= 5.35) took part in this study, all of whom were right-handed. Pre-trial screening questionnaires ensured that none of the participants suffered from impairments of their motor ability or an oversensitivity to cyber sickness, which has often been associated with HMD use.

2.5.2. Experimental apparatus

This experiment was conducted in close collaboration with the Institute of Automatic Control Engineering (LSR) of the Technische Universität München, which provided the experimental apparatus and implemented the assistance functions to be investigated.

Operator Environment

Two 7-DOF ViSHaRD7 robotic arms were used as haptic interfaces (see Figure 33).

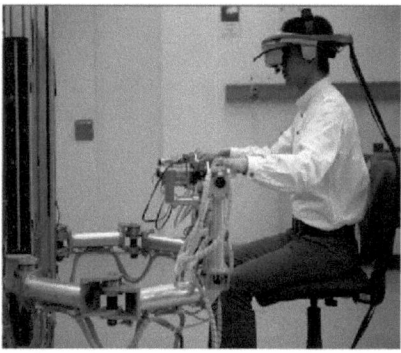

Figure 33. Operator setup with ViSHaRD7 and Head-Mounted Display.

The arms have a relatively large, anthropometric workspace and a high force output capability. The redundancy is used to decouple translational from rotational movements. Thus, movements were easily restricted to translational movements in the horizontal plane. JR3 6-DOF force/torque sensors are mounted at the end-effectors of the two devices and end-effector positions are obtained by applying the forward kinematics to the measured joint angles. There was no division of movement between the two arms, i.e. each arm could move the box in all directions along the x/y plane.

Teleoperator Environment

A virtual reality scenario depicting a maze was presented to the participants through a head-mounted display (HMD) with SXGA resolution and a frame rate of 30 Hz (see Figure 34). In this virtual maze, a box was placed which the operator controlled via two virtual proxies.

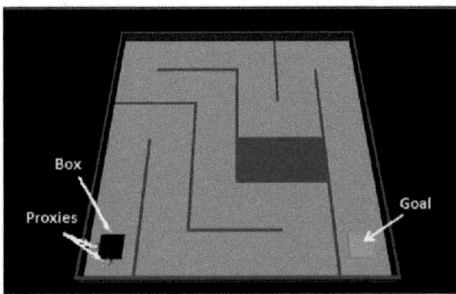

Figure 34. Screenshot of the virtual maze. The (blue) box is moved by proxies permanently attached to the box. The goal position is marked by a (green) square on the right.

2.5.3. The assistance functions

Three different assistance functions were implemented which aimed at assisting the user in moving the virtual box through the maze and were applied to the box itself, rather than the user's virtual proxies. The assistance functions were designed to offer different levels of autonomy over the operator's control. Hence, one task-independent and two task-dependent assistance functions were selected and slightly modified for the experiment, as described in the following:

Assistance function 1: Variable damping (VD)

As a task-independent assistance function, which affords the user a high degree of autonomy over movements, a variable damping approach similar to that proposed by Duchaine & Gosselin (2007) was chosen. With this approach, accelerating motions are facilitated with decreased damping, while decelerating motions are assisted with increased damping.

Assistance function 2: Passive/active virtual fixture combination (VFC)

A combination of passive and active virtual fixtures was implemented for this assistance function. This combinatory assistance concept passively guides the box along the path, as long as it is moving towards the goal position. Movements that are not conducive to fulfilling the task objective, i.e. deviations from the pre-specified path, are constrained actively. The desired path was specified as the centreline of the maze, with circular movements for the steering manoeuvres around corners. For the experiment, a scaling of 1.5/0.9 was applied to the forces tangential to the path in the desired/undesired direction. The orthogonal forces constraining deviations from the path were computed using a virtual spring with a stiffness of 8000 N/m. Since this task-dependent assistance function actively constraints the user, it somewhat curtails the user's autonomy over the control of the box.

Assistance function 3: Active virtual fixture (AVF)

The third haptic assistance function to be investigated constituted a strong, active virtual fixture, which pulled the user tangentially along the pre-specified path with a constant force of eight N. The virtual spring orthogonal to the path was set to 720.000 N/m. Even if the operator did not move the box, the fixture would pull it towards the goal position. Thus, this assistance function considerably curtails the user's autonomy over the movements of the virtual box.

For further technical specifications of the experimental apparatus and the selected assistance functions, the reader is referred to Nitsch, Passenberg, Peer, Buss & Färber (2010) and Peer & Buss (2008).

2.5.4. Experimental design

A repeated-measures experimental design investigated the effects of three assistance functions (VD/VFC/AVF) and a control condition with no assistance (NA) on task performance and users' perceived level of comfort in working with the different assistance functions. Task completion time (sec.) and the accumulated time that one of the virtual proxies or the box spent in contact with a wall (total collision time) (msec.) constituted objective measures of task performance. In addition, the perceived level of control over movements of the virtual proxies was assessed on a percentage scale. Subjective assessments of task performance and the users' perceived comfort in working with the different assistance functions were assessed in pairwise comparisons of the different assistance function conditions. The pairwise comparison featured a fully balanced, randomised design, whereby each stimulus pair was tested in both directions: Condition1-Condition2 and Condition2-

Condition1. Furthermore, it allowed for three response alternatives: "better than", "worse than", "equal than".

2.5.5. Procedure

Participants were informed of the risks of cyber sickness and signed consent was obtained. Prior to the experimental trials, participants were given the opportunity to familiarise themselves with the experimental setup and the task requirements, which were to move the virtual box in the maze from a pre-defined starting point to a goal location using their input devices. Using standardised instructions, particular emphasis was placed on the requirement to avoid contact with the walls, whilst moving as quickly as possible through the maze. Prior to the experimental run, participants conducted a number of practice trials, during which the collision time and the task completion time were measured and compared to respective time standards, which had been established during an extensive pilot testing phase prior to the user tests. After each practice trial, they were given feedback on whether they had performed too slowly, or had produced too many collisions, or both. When the participant was able to perform the task twice within acceptable margins of the specified standard on both criteria, the practice phase was completed and the experimental trials commenced. This instruction phase was designed specifically to ensure that all participants would approach the task in a somewhat consistent manner, so that effects could be attributed confidently to the experimental manipulations rather than individual style or ability. Upon completion of the practice phase, each participant performed the practiced task a total of twelve times. Each trial was conducted with a different assistance function or the control condition, whereby each type of assistance (including the control condition) would be implemented a total of three times. The order, in which the different types of assistance were implemented, was systematically varied. After each trial, participants completed a short questionnaire.

2.6. Results

Due to missing data, one person's data set had to be excluded from the statistical analysis of the results, leaving N=31. Data were inspected for outliers and tested against the assumptions for parametric tests (see Appendix A). Where the assumptions were met, data with z-scores of greater than 3.29 were considered outliers and consequently excluded from further analysis.

2.6.1. Perceived control over TOP movements

H1: Task-dependent assistance functions are necessarily more imposing than task-independent assistance functions (Abbott, Marayong, & Okamura, 2007). Hence, the four conditions of assistance will differ in users' perceptions of their control over TOP movements, with task-dependent assistance functions being perceived as taking more control away from the user than task-independent assistance functions.

Perceived level of control was measured after each trial with a single-item scale on which participants indicated, how much control they felt they had over the movements of the box in a specific trial (0% - 100% in control). In order to ascertain whether the perceived amount of control over movements has an effect on objective and subjective task performance and user perception, it was determined whether the user rating of perceived control constituted a reliable measure and whether users actually perceived a difference in the amount of movement control provided by each assistance function.

First, in order to assess whether the ratings to the "Perceived level of control" scale were reliable, the intra-class correlation coefficient (ICC) was calculated (N=31), using a random effects model. The analysis yielded a Cronbach's[12] α value of .99, which, according to Kline (1999), indicates very high reliability of participants' ratings.

Second, a repeated-measures ANOVA was conducted with the control ratings in order to determine whether the assistance functions varied in the amount of movement control they provided. The ANOVA found no significant repetition effect on the perceived level of teleoperator movement control ($F(2,58) = 1.70$, $p = .19$), indicating that level of control ratings for each assistance function did not vary significantly over the three measurement intervals. In other words, practice with each assistance function did not affect participants' impressions of movement control. On the other hand, the ANOVA did reveal a significant assistance function main effect ($F(1.91, 55.37) = 92.68$, $p<.001$, $\eta_p^2 = .76$, $\varepsilon = .64$), indicating that the assistance functions varied significantly in the amount of control they afforded the participants over TOP movements. Mean values and standard deviations for the control ratings of each assistance function are displayed in Figure 35.

---

[12] Cronbach's α indicates the lower bound for the true reliability of a questionnaire. It is defined as the proportion of the variability in the responses that is the result of differences in the respondent and is computed based on the number of items in a questionnaire and the ratio of the average inter-item covariance to the average item variance (Cronbach, 1951). Kline (1999) proposes that values > .7 indicate a reliable measure.

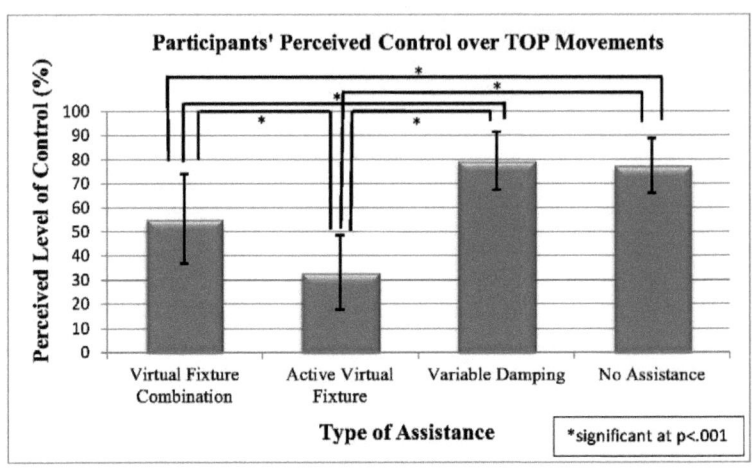

Figure 35. Mean values and standard deviations for estimates of control over TOP movement for each assistance function condition.

Bonferroni-adjusted post-hoc comparisons indicate significant differences in mean ratings between all assistance function conditions except between control ratings of variable damping assistance and the control condition (see Appendix D., Table 14). The means indicate that participants felt they had least control over TOP movements with the (task-dependent) active virtual fixture. In fact, they felt that the system was largely responsible for TOP movements. The other tested task-dependent assistance function, which combined a passive virtual fixture with a weak active fixture, seemed to approximate an equal cooperation between the user and the system as participants felt they shared control with system on almost equal terms. On average, participants indicated that they had slightly more movement control with variable damping, the task-independent assistance function, than in the control condition, although the difference in ratings between these two conditions was not significant. Note that, although participants felt they were largely in control of movements in the control condition, participants did not feel as though they were completely in control, even though technically they were. Most likely, this finding reflects a known tendency of people to avoid extreme answers on rating scales (Bortz & Döring, 2006). The results clearly indicate that the users felt that task-dependent assistance functions took more control away from them than the task-independent assistance function, thus supporting hypothesis H1.

2.6.2. Objective task performance measures

H2: Provided that the task and the environment can be precisely pre-defined, task-dependent assistance functions can provide greater assistance than task-independent assistance functions (Abbott, Marayong, & Okamura, 2007). Since this condition applies in the present experiment, assistance *functions will lead to greater improvement in objective measures of task performance, the less control is provided to the user.*

In order to gauge the impact of the different assistance functions on objective task performance measures, parametric, repeated-measures ANOVA were conducted with type of assistance (VFC, AVF, VD, NA) as an independent variable and collision times (msec.) and task completion times (sec.) as dependent variables. Where Mauchley's test of sphericity was significant, degrees of freedom were adjusted with a Greenhouse-Geißer Correction, and ε values are reported.

Total collision time

Practice was not found to make a difference with regard to collision avoidance ($F(1.60, 47.94) = 0.98$, $p = .38$, $\varepsilon = .79$). However, the ANOVA did find a significant main effect of assistance on total collision time ($F(1.72, 51.72) = 34.67$, $p < .001$, $\eta_p^2 = .54$, $\varepsilon = .58$), indicating that the time that participants spent in contact with a wall varied significantly between the different assistance conditions. Mean values and standard deviations are displayed in Figure 36.

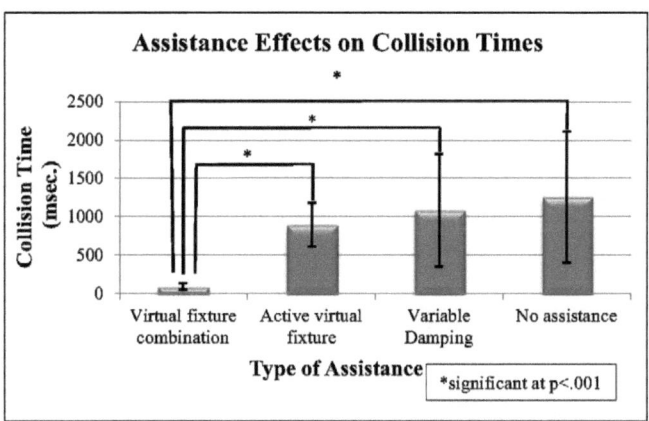

Figure 36. Mean times spent in contact with a wall and corresponding standard deviations for each assistance function and the control condition.

Bonferroni-adjusted post-hoc comparisons revealed that, with the virtual fixture combination, collision times were significantly shorter compared to performance assisted by the active virtual fixture, variable damping or the control condition. On the other hand, the other three conditions did not differ significantly in collision times (see Appendix D., Table 15). In other words, only the virtual fixture combination significantly reduced wall collisions.

Task completion time

Despite the rigorous pre-experimental training phase, practice had a significant effect on task completion times ($F(2,60) = 33.73$, $p<.05$, $\eta_p^2 = .53$). The means indicate that participants were on average faster on the third trial of each assistance condition than they were on the first trial. Of greater relevance to the hypothesis is the finding of a significant assistance function main effect on task completion times ($F(3,90) = 36.74$, $p<.001$, $\eta_p^2 = .53$), which was found to be independent of the amount of practice that participants received, as no significant interaction effect was found ($F(6,180) = 0.32$, $p = .93$). Figure 37 shows a bar chart, which plots mean task completion times for each trial and assistance function condition.

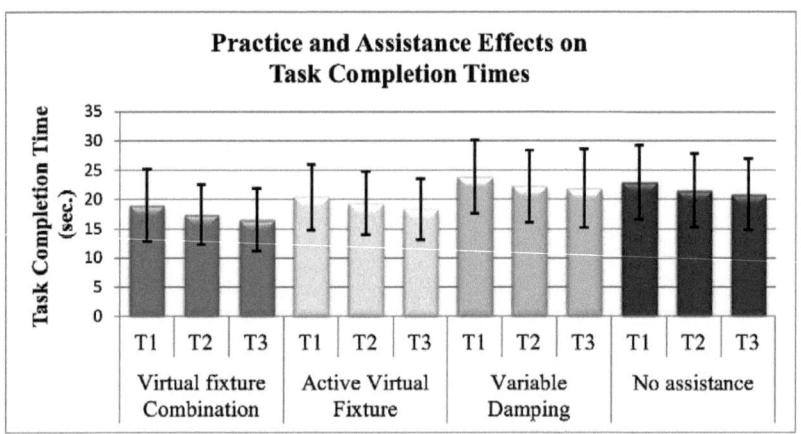

Figure 37. Mean task completion times (sec.) and standard deviations for each trial and assistance function.

Bonferroni-adjusted post-hoc comparisons of average task completion times between each assistance condition showed significant differences between all conditions, as indicated in Appendix D., Table 16.

The results suggest that performance with the virtual fixture combination led to the fastest task completion times ($M_{vfc} = 17.68$ sec., $SD_{vfc} = 5.36$ sec.), followed by the strong active virtual fixture ($M_{avf} = 19.40$ sec., $SD_{avf} = 5.17$ sec.) and the control condition ($M_{na} = 21.84$ sec., $SD_{na} = 6.01$ sec.). On average, the variable damping produced the slowest task completion times ($M_{vd} = 22.68$ sec., $SD_{vd} = 6.19$ sec.).

To summarise, the assistance function that offered the least amount of control to the user (i.e. the active virtual fixture) did not lead to the best task performance. Instead, the combination of passive and active virtual fixtures seemed to have offered the most assistance to the user in performing their task. Therefore, hypothesis H2 is not supported by the data.

2.6.3. Comfort assessments

H3: People feel uncomfortable when their control over their movements is curtailed (Lefcourt (1973); Scepkowski & Cronin-Golomb (2003); Syrdal, Koay, Walters, & Dautenhahn (2007)). Hence, it is predicted that the level of comfort will be determined by the perceived amount of control that the assistance functions afford the user. Specifically, it is expected that users will feel less comfortable working with an assistance function that curtails their movement autonomy strongly, and feel more comfortable when performing with more movement control.

Perceived level of comfort in working with each assistance function had been assessed with a questionnaire item in pairwise comparisons of the different assistance functions. The pairwise comparisons were analysed according to Case V of Thurstone's Law of Comparative Judgement (LCJ) (Thurstone, 1927). The LCJ assumes that observers are reliable and that the stimuli are judged based on the same criterion. Accordingly, the data were checked for intransitive judgements[13] and inter-rater agreement[14] prior to analysis, both of which met the assumptions of the LCJ. The application of this measurement model allows for the transformation of comparative preference data to a single group composite interval-type scale, provided that the answers relate to a uni-dimensional continuum. Essentially, this scale is constructed based on the estimated probability that one item will be preferred to another, which can be converted into z-values. The mean of these z-values constitute the individual attribute's scale values. In order to avoid negative scale values, a linear transformation has been conducted in which the absolute value of the lowest score was added to each score. The resulting interval-type scale is presented in Table 1.

Table 1. Scale values for ratings of perceived comfort in working with each type of assistance.

Type of Assistance	Active Virtual Fixture	Virtual Fixture Combination	No Assistance (Control)	Variable Damping
Scale Value	0	1.31	1.40	1.51

---

[13] Intransitive judgements indicate poor consistency of judgements for individual raters. E.g. if condition a is preferred to condition b (a > b) and condition b is preferred to condition c (b > c), condition a should also be preferred to condition c (a > c). If a < c, the judgement is intransitive.

[14] Inter-rater agreement is assessed with Kendall's coefficient of concordance (W) which designates the degree of association between *k* sets of rankings.

The scale values indicate that users felt least comfortable when performing their task with the active virtual fixture, presumably since the active virtual fixture offers the user the least amount of control over movements. On the other hand, users preferred the control condition and the variable damping, which did not perceptibly reduce control over TOP movements. These findings seem to offer support for the stated hypothesis H3.

2.6.4. Subjective task performance assessments

H4: The perceived level of control provided over TOP movements can influence the way people perceive their performance with an assistance function (Kim & Hinds (2006); Hogg & Vaughan (2002)). It is therefore predicted that performance judgements will deteriorate with decreasing levels of control provided to the user.

Judgements of perceived task performance were also analysed according to Thurstone's Law of Comparative Judgement. The resulting scale is presented in Table 2.

Table 2. Scale values for ratings of perceived task performance quality.

Type of Assistance	Active Virtual Fixture	No Assistance (Control)	Variable Damping	Virtual Fixture Combination
Scale Value	0	1.25	1.29	1.45

The scale implies that participants thought the active virtual fixture resulted in the worst performance, followed by variable damping which almost ties with the control condition. The combination of active/passive fixtures was preferred the most when judging the quality of task performance. Yet, the hypothesis would predict that the control condition and the variable damping would be perceived as showing the best performance. Hence, this finding does not directly support hypothesis H4 but rather suggests an interaction between actual performance (for which the virtual fixture combination was found to be most effective) and the level of control provided.

2.7. Discussion

Previous studies evaluating haptic assistance functions focused on their effect on objective measures of task performance. In response to these studies, it was argued that objective task performance measures are not sufficient in demonstrating the value of assistance functions for the industrial use in teleoperation systems. Since haptic signals are more intrusive than audio or video signals as they curtail the user's autonomy, their effect on the user's perception of task performance and their level of comfort in working with such an assistance function should also be considered. The purpose of the present study was therefore to investigate the influence of three different haptic assistance functions that offer different levels of control over TOP movements on task performance and user perception. Of particular interest was the role of perceived movement control in judging task performance and the user's level of comfort during task performance. Furthermore, it was also investigated whether a task-independent assistance function can be similarly effective in improving movement coordination as task-dependent assistance functions.

The experimental results suggest that the strong, active virtual fixture was perceived as offering the user the least amount of control over movements. Against expectations was the finding that this assistance function did not lead to the greatest task performance improvements. Although performance with this type of assistance was significantly faster compared to unassisted performance, it did not significantly reduce collision times. In contrast, an assistance function utilising a combination of active and passive fixtures was found to improve performance significantly in terms of speed and collision avoidance, even though technically, it offered less support to the user. Two explanations for this finding come to mind. First of all, it is conceivable that the strong active virtual fixture simply did not provide adequate support for users' movements. However, this seems unlikely, as the active virtual fixture was designed to complete the task on its own without any collisions whatsoever. Instead, it would seem more likely that task performance would suffer from the strong active virtual fixture as participants struggle to gain control over the movements and thereby actively work against the provided assistance. In contrast, the weaker, combined assistance function offered support to the user but was not noticeable enough to provoke compensatory, task-aversive behaviour. Overall, it also seems that the task-dependent virtual fixtures offered better support for task performance than the investigated task-independent assistance function, as the variable damping only seemed to slow down the user, without significant support for the avoidance of collisions. It seems that in cases where the task demands are unstructured and maximum flexibility is required, e.g. during remote exploration of unknown territory, little assistance can be provided. Future studies are needed to investigate whether this concept can be improved upon or whether other sensory modalities (e.g. in the form of visual cues) can be utilised in order to effectively compliment or substitute task-independent haptic assistance functions.

Finally, the data indicate that when participants felt that they were not in control of the movements of their virtual proxy in a particular trial, they were more likely to consider negative task performance aspects in their judgment of performance. That is, collisions in trials in which the system was perceived to be in control resulted in decreased subjective performance and level of comfort ratings, even though the user collided less with the wall when the assistance function was dominating the movement of the manipulated object. An explanation for this finding can be derived from other empirical findings, e.g. (Kim & Hinds, 2006), as well as the social psychological literature, e.g. (Hogg & Vaughan, 2002). These suggest that the entity that is perceived to be in control is generally also held accountable for mistakes. By the same token, literature on attributional bias suggests that users tend to attribute their own performance mistakes to cross-situationally varying factors, whereas they attribute the system's mistakes to inherent flaws of the system rather than coincidence or the difficulty of the task. The findings of the present study thus seem to suggest a certain trade-off between control and subjective evaluations: if an assistance function cannot keep performance errors to an absolute minimum, it is more likely to be judged unfavourable, the more intrusive it is. On the other hand, if assistance functions provide the user with more than 50% of movement control, the user is more likely to overlook performance errors made.

In conclusion, the experimental evaluation of the implemented haptic assistance functions shows that some assistance functions significantly improve movement coordination compared to unassisted task execution. Although task performance was overall better with task-dependent assistance than the task-independent function, it was found that it is most likely not the aspect of task dependence, but the level of perceived movement control that primarily influences the user's perception of the quality of their performance. It seems that the more the operator feels in control of the movements, the more likely it is that he or she will overlook performance errors in judging task performance. On the other hand, once the operator does not feel in control, evaluations are likely more influenced by mistakes, even if, in fact, fewer errors were made. Users also felt uncomfortable when they did not feel in control of their movements. In industry, this could mean that users would either avoid the use of this system or find a way to circumvent this safety precaution (Glendon, Clarke, & McKenna, 2006). It would also seem likely that the level of trust which users place in an assistive system would affect whether or not users feel comfortable with and/or work against strong assistance, an assumption which warrants further investigation. Overall, the findings of this study indicate that user perception is an important factor in the evaluation, and consequently applicability, of assistance functions in teleoperation systems.

How much autonomy an assistance function should allow the user in order to achieve the best possible task performance, is certainly strongly dependent on the specific demands of the task in question. Yet, based on the presented results, one might argue that for optimal task performance, haptic assistance should be as strong as necessary and allow for as much autonomy as possible. Although it is technically difficult to realise, the solution seems obvious: as Wickens (1992) has pointed out, the ideal human-machine system would be adaptive in that the user may choose the appropriate level of assistance for any given task and environment.

3. Guidelines for the implementation and effective use of haptic guidance

The experiments investigating task-facilitative effects of haptic guidance signals demonstrate clearly the importance of the correct application of these signals. While virtually no effect of haptic expert demonstrations prior to the task on task performance could be ascertained, haptic assistance functions that guide the user during task performance have been found to be very helpful in improving performance. A review of the available literature and the previously discussed experimental results allowed for the derivation of a number of guidelines for the effective use of haptic guidance signals in teleoperation systems that should be considered in order to capitalise upon their effects.

- In cases where a correct replication of forces is important to achieving task goals, haptic demonstrations of the desired level of force application may be helpful.
- When it is important to learn specific or general motor skills, rather than demonstrating a task to a novice haptically, it is recommended to promote active learning through practice and employ haptic feedback from the remote environment during training. Feedback of mistakes and correct responses should also be provided during training.
- For tasks that require flexibility, task-independent assistance functions, for instance with variable impedance control, may be used to render work more comfortable, though they do not necessarily improve performance with the system.
- If the task demands and the environment are known beforehand, task-dependent assistance functions are recommended in order to prevent undesired actions of the user and reduce time demands. Care should be taken that not too much autonomy is taken away from the user, as it may make the user feel uncomfortable, make mistakes more noticeable and may even be dangerous in the event of system failure or flawed virtual models. A combination of passive and weak active virtual fixtures, which leaves around 50% of (perceived) movement control to the user, was previously found to be most effective in terms of performance, whilst still being comfortable to work with. Preferably, an adaptable level of support by virtual fixtures would be provided, which may be adjusted by the user according to user and task demands.

Chapter IV.

Meta-analytic investigations into task facilitative effects of haptic signals

The empirical investigations detailed in previous chapters served to fill gaps in the literature and to provide guidelines for the effective employment of haptic signals in specific task domains (i.e. pick-and-place, abrasion, tracking, transporting). The results have partly confirmed previous research findings of similar studies that used different experimental setups (e.g. regarding the effectiveness of force feedback in reducing excessive forces), while in others, they flatly contradicted previous research (e.g. regarding the ineffectiveness of haptic expert demonstrations in improving motor performance). Since the work domain (e.g. surgery, micro- or macro-assembly), the task domain, the experimental methodology and the experimental apparatus vary for virtually every study that has been conducted in the field of haptic teleoperation applications, it is difficult to ascertain the overall effectiveness of a particular haptic application.

Furthermore, upon review of the experimental literature, one cannot help but gain the impression that in their conclusions, researchers are quick to generalise positive effects of their findings beyond the scope of their study, while a negative effect or a lack of effect is usually attributed to the experimental set-up or methodology used in the particular study. Since, as in most investigative sciences, studies are rarely replicated and a lack of effects is infrequently reported, the reader is left wondering whether haptic feedback is generally effective in improving task performance and what the extent of its effects may be. The latter point is especially important when it comes to deciding whether an investment into haptic interfaces, be it in terms of money or research effort, is worth making.

In order to provide reliable and comprehensive advice on the effective use of the investigated applications of haptic signals, it is therefore vital to establish to what extent the findings of the studies presented in this work generalise to other haptic devices, as well as different task and work domains, and are not confined to the specific experimental setups of the presented studies. In the absence of an overarching theoretical framework that could potentially substantiate empirical research, a synthesis of available research evidence is required, so that it may be determined whether a particular application of haptic signals is found to be effective (or ineffective) in a variety of settings, rather than in one particular experimental setup.

1. Previous efforts to synthesise research on haptic interfaces

An in-depth review of presently available empirical literature suggests that haptic interface research is a rather fragmented field. It seems that, up to now, only few attempts have been made to present the many bits of existing research from a broader, more integrative perspective. Tan, Eberman, Srinivasan & Cheng (1994), and MacLean (2000) independently reviewed haptic interface technology available at the time. In their reviews, the authors examined and categorised physical characteristics and demands of haptic interfaces and offered guidelines for their effective design. Neither of these reviews, however, disseminated available empirical evidence on task performance with these systems. As such, the suggested guidelines focused on physical characteristics of the interface design, rather than effective employment of existing haptic technology.

Only few reviews considered the effect of haptic signals on measures of human performance with a technological system. Among them, Hale & Stanney (2004) disseminated psychophysical, neurological and physiological studies that involved haptic displays. Based on their selected studies, they proposed guidelines for the design of haptic and multimodal displays and theorised regarding possible beneficial effects of various forms of tactile and kinaesthetic feedback on texture perception, 2D/3D form perception and spatial awareness. Since teleoperation systems were not considered, however, explicit measures of force application and movement coordination with these systems did not feature in their review.

Jones & Sarter (2008) presented a detailed review of vibrotactile devices. In their review, the authors formulated guidelines that advise on the most effective employment of vibrotactile devices based on their consideration of previous studies on human vibrotactile perception, general mechanical properties of vibrotactile display technology, and human performance with these systems. Yet, similarly to Hale & Stanney (2004), the authors considered neither force feedback devices nor relevant measures of the effect of tactile displays on human force application.

Crucially, none of the previously mentioned reviews considered the strength of effects that were found in the various pieces of research that the authors disseminated, nor did they produce indicators of overall effect strength of particular applications. Yet, the consideration of the effect strength of individual findings, as well as the indication of an overall effect strength of an application is important for two reasons. For one, when considering multiple studies on the same topic that differ in their results, effect strength can serve to judge which findings are more reliable and which ones are less so. Furthermore, without an indication of the overall effect strength of a particular application, it is difficult to judge whether this application can be recommended for the industrial employment of haptic applications, considering that marginal improvements of human performance may not necessarily be worthwhile a substantial financial investment.

Burke, Prewett, Gray, et al. (2006) were among the few researchers who considered the strength of effect that haptic signals exert on human performance. Specifically, the authors attempted to summarise and evaluate empirical evidence on general effects of visual-tactile feedback on certain performance measures such as reaction time and error rate. In a meta-analysis of 43 studies, they demonstrated that the multi-modality of visual-tactile feedback provides a significant, albeit tentative, advantage over using an exclusively visual feedback system, whereby the extra modality was found to improve reaction time, but not error rate. However, the authors did not differentiate between different applications of haptic technology and focused mostly on studies that investigated two-dimensional human-computer interaction. As such, their findings can only marginally advise on the effective use of particular haptic applications in teleoperation systems, which typically entail three-dimensional human-machine interaction.

In summary, although crucial for the development of guidelines that advise on the most effective use of industrial haptic applications, systematic reviews of empirical evidence on task facilitative effects of haptic signals in a teleoperation context are sorely lacking. The following chapter aims to address this gap. In order to determine the nature and strength of effects of haptic signal applications on human performance with teleoperation systems, independent of the experimental setup used in a particular study, a meta-analysis with selected empirical evidence was conducted. Based on the results of this analysis, overall effect strengths of the investigated haptic applications are provided, which may be used to calculate the financial gain that may be achieved with the use of haptic displays in industry. Moreover, the analysis aims to highlight the strengths and weaknesses of the investigated haptic applications from which guidelines on the use of these applications are derived, which have proven effective in a variety of experimental settings.

2. Meta-analysis as an investigative tool

Meta-analyses allow for a meaningful numerical comparison and analysis of studies that use different measures and manipulations (Lipsey & Wilson, 2001). In essence, a meta-analysis combines standardised effect sizes over a number of studies to compute a summary effect that represents the weighted mean of the individual effects, which were found in each of the included studies. The weights that are assigned to the individual effect sizes, and thus the emphasis that is placed on each of the included studies, are determined based on assumptions about the distribution of effect sizes from which the studies were sampled, and are calculated using established procedures (Borenstein, Hedges, Higgins, & Rothstein, 2009). In simple terms, the summary effect is the mean of the effect sizes of the included studies, with more weight assigned to the more precise studies. This summary effect size statistic may be used to determine the statistical significance of an overall effect, as well as indicate both the direction and the magnitude of a relationship between an independent variable and an outcome measure. Further insight into the meaningfulness of an effect can be obtained from indicators of the accuracy of this summary effect as an estimate of the true effect, as well as measures of the consistency of a particular effect across all sampled studies (Borenstein, Hedges, Higgins, & Rothstein, 2009).

Meta-analysis offers a number of advantages over traditional, narrative literature reviews, which focus on the reported statistics of individual studies. Since a meta-analysis typically aggregates the effect sizes of a larger number of studies, it has the advantage that it is relatively unaffected by small sample size and other methodological confounds that are often associated with individual studies. This makes meta-analysis results much more robust than individual studies. Moreover, in contrast to narrative literature reviews, where the authors' criteria for selecting and emphasising particular studies over others are usually implicit, meta-analysis aims to use transparent criteria so that other researchers may reproduce it. Hence, meta-analysis offers the benefit of providing a quantitative and transparent basis for the evaluation of current practices, which makes meta-analyses especially attractive for the formulation of practice guidelines (Lipsey & Wilson, 2001).

3. Data selection criteria

Meta-analysis has often been termed a Garbage-In-Garbage-Out (GIGA) technique, which means to imply that the results of this analysis are only as useful as the data that is fed into it. In the selection of data that provides input for the meta-analysis, the researcher is forced to make a number of decisions, all of which have consequences for the meaningfulness of the results. It is therefore vital that the underlying decision processes are made transparent. Hence, in the following section, general selection criteria that were considered during the literature review process of the present meta-analysis are described, before details of the analysis itself are described in the Method Section, p. 116 ff.

3.1. Research quality

For one, the level of research quality of each study needs to be considered, before it is included in a meta-analysis. Generally speaking, the lower the quality of a particular study (e.g. due to methodological flaws), the less reliable the result of that study will be. While meta-analysis is fairly robust to the effects of methodological flaws of a few studies, obviously, the meaningfulness of the results of a meta-analysis will diminish if these are based on a large number of flawed studies. On the other hand, if the quality criteria are too stringent, only few studies are likely to meet them and thus be eligible for inclusion. Yet, a number of experts in the field have argued that the inclusion of a greater number of studies of reasonable quality is preferable to the selection of few studies that are of high quality, as long as the criteria for selection are specified (Bortz & Döring, 2006, p. 675). The challenge is thus to set an acceptable level of research quality, which applies to a larger number of studies.

This challenge is exacerbated by the generally low quality of empirical research on human performance with haptic technology. While the subject is inherently interdisciplinary in nature, the majority of studies is conducted in intradisciplinary teams, which typically lack the required level of professional training in other disciplines. As a result, objective and reliable data, whilst certainly aimed for, are infrequent. Flaws in the design, execution, interpretation and the reporting of human user studies are widely prevalent. For example, since prior research on haptic interfaces has largely been technology-driven, with an emphasis on technological feasibility rather than need or purpose, studies often lack any specifications of testable hypotheses or fail to rationalise these. HARKing (Hypothesising After the Results) is also common practice, with a rather unconcealed evaluative bias in favour of newly developed technologies. Frequently left out are mention of randomisation technique (if any), standardised instructions or the use of a practice/familiarisation phase, making it very difficult to confirm effects with replication studies. Quite a few studies reported that data of participants were excluded due to their failure to "properly follow instructions" or unexpected reactions. Wide generalisations of findings are often founded on exceedingly small sample sizes, with the majority of studies testing less than ten participants.

Furthermore, dependent variables are often poorly operationalised, with many studies reporting anecdotes rather than quantitative or qualitative data. Where quantitative data is reported, statistical analysis is often lacking or flawed. Few studies conducted inferential statistics, most of those that do only report them selectively, thus producing a pronounced outcome reporting bias. In many cases, the choice of statistical model tested can hardly be justified and the treatment of statistical outliers is almost never reported. Finally, whilst probability values are oftentimes supplied by those studies that conduct inferential statistical tests, these tend to be restricted to effects that were found to be statistically significant. Human user studies that lack the required methodological rigour and analytical expertise, whilst not without merits, cannot provide conclusive scientific evidence for a particular effect, as they lack both reliability and validity. Therefore, for the present meta-analysis, it was decided that studies would not be included in the meta-analysis if they are suspected, based on the above-mentioned considerations, to present unreliable effects.

3.2. Construct definitions

When it comes to the selection of studies, further decisions need to be made in the definition of the investigated constructs, i.e. independent and dependent variables. For example, a decision needs to be made whether all studies that feature any form of haptic feedback are eligible for inclusion or whether only those that feature a specific haptic device or a particular sample (e.g. expert users) should be included. Similarly, do studies qualify for inclusion if they investigate the effect of haptic signals on any type of performance measure or only on specific criteria? While the primary aim of meta-analysis lies in the synthesis of a wide range of studies that differ in their scope and methodology, critics have argued that an overly heterogeneous selection of studies further curtails the meaningfulness of results (the apples-and-oranges argument) (Bortz & Döring, 2006, p. 675). The challenge in this regard is therefore to decide on an acceptable level of heterogeneity in the sample of selected studies, so that a range of studies with different setups are included in the meta-analysis, but a synthesis of their results is still conceptually meaningful. The following section hence specifies the construct definitions that are applied to the present meta-analysis.

3.2.1. Haptic applications

It was decided that studies are considered for inclusion if they feature one of four applications of haptic signals and were contrasted with a control condition in which the respective application was absent. The applications of interest were (a) vibrotactile feedback, (b) force feedback, (c) haptic assistance, and (d) haptic expert demonstrations.

a) The category vibrotactile feedback encompasses studies that made use of any devices that provide directed or diffuse vibrations in order to alert or inform the user of contact or direction. The literature review suggested that applications for vibrotactile feedback seem to have focused largely on two-dimensional human-computer interaction. However, since the focus of the present work is on human-machine interaction in teleoperation systems, it was decided that only studies that investigated this haptic application within a teleoperation context would be taken into consideration.
b) Into the force feedback category fall studies which employed admittance- or impedance-type force feedback devices of any output capability that simulate contact in the virtual or remote environment.
c) The haptic assistance category constitutes studies that employ force feedback in order to assist the user task-dependently or task-independently, actively or passively, through guiding or by preventing the user from entering forbidden regions.
d) Finally, the haptic expert demonstration category includes studies that aimed at training users' movement coordination and/or force application skills by exposing them to active haptic guidance, which had been previously recorded by skilled experts.

3.2.2. Task performance

It was further decided that only studies would be considered for inclusion if they featured task performance outcomes that encompass force application, task completion time and/or error rate. In this context, the measurement of force application pertains to all measures of users' accuracy in the application of forces, such as peak forces, mean forces or force variance. The variable task completion time refers to the measured times for entire tasks as well as that of individual task segments. Finally, the error rate captures any deviation from a response previously defined as correct. The error rate includes, for example, position or force deviation measures as well as failure/success rates of tasks that required quick or precise movements. In contrast, studies that only featured errors in qualitative judgements (e.g. weight or texture discrimination) would not be eligible for inclusion.

3.3. Type of effect

An important criterion for meta-analysis is the independence of the individual effect sizes that are included. That is, for each study that is eligible for inclusion, only one effect statistic should be included in the meta-analysis. Since most studies report statistics on a number of analyses, another decision hence needs to be made on the inclusion of specific effect sizes. While typically, meta-analyses are conducted in order to determine average mean effect strengths of a particular variable, it was decided that the present analyses should aim at determining average maximum effect strengths instead. This means, that if several effect sizes are presented for a particular outcome measure, the largest effect, rather than a mean calculation of all reported effects, should be included in the analysis. This choice was made based on two reasons. For one, effect sizes can only be accurately determined if the appropriate statistics are reported in the literature. Yet, unfortunately, it has become common practice to report only statistics for significant results, while non-significant findings, if reported at all, are typically only mentioned as such in the text. In these cases, effect sizes would be calculated based on the assumption that $p = .05$, thus leading to inflated effect sizes and consequently an overestimation of the average effect of haptic signals. Secondly, since measures of effect strength lend themselves particularly well to cost-benefit analyses (Lipsey & Wilson, 2001), the results of the meta-analysis featured in this work may be used to judge the financial impact of a particular investment. It may be reasoned that if the maximum effect of an application is not found to be effective in improving task performance, further investigations and investments to this aim are surely not worth making. Thus, not only can the maximum effect strength be more accurately estimated than the average effect strength, one might also argue that it makes economic sense to determine the maximum return on investments (ROI) into new technologies before deciding whether it is worthwhile to make an investment.

4. Method

Once the basic selection criteria are decided upon, the meta-analysis can be conducted. The following section describes the method in detail.

4.1. Literature search

Multiple approaches were taken to the literature search. For one, relevant key terms (e.g. force feedback, vibrotactile feedback, haptic virtual fixtures, vibrations, user study, haptic display, and haptic device, haptic interface, haptic assistance) were entered into relevant databases and search engines such as PsychInfo, Cambridge Scientific Abstracts (CSA), the IEEE Explorer and Google Scholar. With the exception of the four studies presented in the previous chapters of this work, only published articles were considered. In addition, a 10-year retrograde hand-search was conducted on the table of contents for the journals "Presence: Teleoperators and Virtual Environments" and "Human Factors", two of the most prolific contributors to this field. This literature search yielded 209 studies, which were then checked for eligibility of inclusion.

4.2. Inclusion criteria

Studies were only included in the meta-analysis if they did not feature obvious methodological or analytical flaws (see p. 112 f.) that might indicate that the featured results are not reliable, and if they met all of the following inclusion criteria. For one, studies were included if they feature a force feedback or vibrotactile display and compared the effects of haptic signals to a control condition featuring no haptic signals with an experimental user study. In order to qualify for inclusion, studies also needed to investigate teleoperation or virtual scenarios, which required accurate movement coordination or force application, and which measured forces, task completion time or error rates. Tasks in which qualitative judgements needed to be made such as texture or weight discrimination were not considered for analysis. Studies with sample sizes of less than four participants were not included. For studies, which otherwise qualified for inclusion but lacked the statistics that are necessary to calculate effect sizes, such as mean values and standard deviations or standard errors, or inferential statistics in general, authors were contacted via e-mail in an attempt to obtain the missing information. If the data could not be retrieved, these studies were excluded also. The results of the studies presented in this work were also included. The criteria specified above yielded 54 studies and 74 effect sizes. Of these, three studies were found to report implausibly large effect sizes ($g > 5.0$) and were hence excluded from further analysis. Thus, the present analysis was conducted with 51 studies and 71 effect sizes, covering a time span of 15 years (1996-2011). A list of all studies included in the meta-analysis can be found in the Reference Section (p. 148 ff.).

4.3. Effect size calculations

All analyses have been conducted based on recommendations by Lipsey & Wilson (2001). Based on their suggestions, the standardised mean-difference effect size (Cohen's d) was determined an appropriate effect size index. The effect sizes were calculated for each included study based on reported summary statistics, such as means and standard deviations, inferential statistics produced by t- and F-tests, or p-values if no other statistics were available. If studies reported effects for more than one outcome measure, e.g. task completion time and error rate, multiple effects were recorded. Oftentimes, studies would report more than one effect for a particular outcome measure, for example, mean forces and peak forces as measures of force regulation performance. Based on the rationale provided in Section 3.3 of this chapter, in these cases, the largest effect sizes were chosen for inclusion in the analysis. Since the vast majority of studies included feature small sample sizes and it was shown that Cohen's d tends to be upwardly biased when based upon sample sizes of less than 20 participants (Hedges & Olkin, 1985), a correction was applied using Hedges' g. All subsequent analyses were conducted with this weighted effect size estimate.

4.4. Analyses

Precision of the effect size estimates is expressed in form of calculated 95% confidence intervals, which indicate the extent to which sampling error is likely to influence the estimate of the effect size. For the analysis, it was assumed that variance in effect sizes stem from study-level sampling error as well as subject-level sampling error. Consequently, a random-effects model was chosen, which typically results in wider confidence levels, thus yielding a more conservative estimate. Homogeneity of effect sizes was calculated with Cochran's Q-statistic, which is calculated as the weighted sum of squared differences between individual study effects and the pooled effect across studies, as well as the I^2 statistic, which describes the percentage of variation across studies that is due to effect heterogeneity rather than chance (Higgins & Thompson, 2002). Unlike other homogeneity measures, such as the frequently reported Q-statistic, I^2 is not inherently dependent upon the number of studies included in the analysis.

4.5. Procedure

In order to ascertain the overall effects of the different applications of haptic signals previously discussed (i.e. vibrotactile feedback, force feedback, haptic assistance and haptic demonstrations), independently of the task, the device or the experimental setup employed, each haptic signals application is analysed separately. First, the overall effect of each feedback on task performance is assessed, whereby, based on the rationale provided in Section 4.3 of this chapter, only the biggest effect size is chosen to represent a study that reports on more than one outcome. In a next step, it was explored whether effects of haptic signals on task performance varied by outcome measure. Inspired by a procedure featured in Burke, Prewett, Gray, et al. (2006), for this outcome analysis only, individual studies could feature in more than one category, if they reported effect sizes for several outcome measures, e.g. force application and error rate. Strictly speaking, this procedure is not theoretically appropriate, as it implies that effect sizes are independent, even though they are clearly not. However, since it is unknown whether the three outcome measures chosen for this study constitute measures of the same underlying dimension of task performance or in fact measure disparate constructs, an analysis separate by outcome was considered necessary. Since it violates the theoretical assumption of data independence, however, this analysis should only be considered exploratory in nature.

5. Results

Based on an analysis of the distribution of standardised mean difference effect sizes for over 300 meta-analyses, Lipsey and Wilson (1993) established benchmarks for the interpretation of effect sizes, which will be used as reference for the following analyses. According to their classification, an effect size of ES ≤ .30 constitutes a small effect, one of ES = .50 a medium effect, and an effect size of ES ≥ .67 classifies as a large effect.

5.1. Vibrotactile feedback

For the sake of clarity and brevity, a summary of the results of meta-analyses on performance-enhancing effects of vibrotactile feedback is presented in Table 3, indicating the number of studies (k) included in each model and the overall number of participants (N). Further indicated are the summary effect size Hedges' g, α-probability values (p), 95% confidence interval estimates (CI) and the homogeneity measure Q. Details of this analysis can be found in Appendix E.

Table 3. Summary of random-effects meta-analysis results for the comparison "vibrotactile feedback present" vs. "vibrotactile feedback absent".

	k	N	g	p	Lower CI	Upper CI	Q
Vibrotactile Feedback	7	131	0.75	<.001	0.42	1.07	6.60, p =.36
Analysis by Outcome	10*	196					
Force Application	5	63	0.29	=.63	-0.88	1.46	23.50, p<.001
Completion Time	3	91	0.79	<.01	0.21	1.36	3.28, p =.19
Error Rate	2	42	0.49	=.12	-0.12	1.11	0.12, p =.73

*Note that several studies may be represented for more than one outcome variable in this analysis.

Even though only few studies were available, the meta-analysis indicated a significant overall effect of vibrotactile feedback on task performance measures ($g = 0.75$, p<.001), with consistent effect size reports ($I^2 = 9\%$). Interestingly, the analysis by outcome indicates that vibrotactile feedback only speeds up task completion times significantly ($g = 0.79$, p<.01), but does not appear to reduce forces ($g = 0.29$, p =.63) or error rate ($g = 0.49$, p = .12) significantly. Inconsistency was only significant for force application measures ($I^2 = 83\%$), although it should be cautioned that homogeneity measures are not particularly reliable if calculated based on three or fewer studies (Lipsey & Wilson, 2001, p. 117).

5.2. Force feedback

A summary of the results of meta-analyses on the effects of force feedback on task performance is presented in Table 4, details are listed in Appendix F.

Table 4. Summary of random-effects meta-analysis results for the comparison "force feedback present" vs. "force feedback absent".

	k	N	g	p	Lower CI	Upper CI	Q
Force Feedback	28	536	1.08	<.001	0.75	1.41	101.61, p <.001
Analysis by Outcome	40*	766					
Force Application	8	128	1.07	<.001	0.46	1.68	21.35, p<.01
Completion Time	20	402	0.70	<.001	0.33	1.06	72.59, p<.001
Error Rate	12	236	0.90	<.001	0.41	1.39	44.63, p<.001

*Note that several studies may be represented for more than one outcome variable in this analysis.

The analysis indicates a large, significant effect of force feedback on performance measures (g = 1.08, p<.001), although, yet again, the significant inconsistency in effect size reports (I^2 = 73.4%) suggests that this effect size may not be particularly representative of the actual effect. An exploratory analysis by outcome variable shows that force feedback demonstrates its biggest strength to lie in the reduction of applied forces (g = 1.07, p<.001), although large effect sizes were also found with respect to error rate reduction (g = 0.9, p<.001) as well as time savings (g = 0.70, p<.001). As it was found to be the case with previous analyses, inconsistencies in the report of effect sizes were significant for all measures (I^2_{force} = 67.2%; I^2_{time} = 73.8%; I^2_{error} = 75.4%). On a side note, it may be of interest to note that 19 out of 28 studies that measured forces used PHANTOM devices by SensAble Technologies, with most of the remaining studies using standard force feedback joysticks and a few testing custom-built force feedback displays. There is a possibility that the effects found in this study are biased by the large group of PHANTOM displays. However, since this group encompasses a number of very different devices, it is unlikely that this bias of further consequence to the interpretation of the results.

5.3. Haptic expert demonstrations

A summary of the results of meta-analyses on the effects of haptic assistance on task performance is presented in Table 5. Details can be found in Appendix G.

Table 5. Summary of random-effects meta-analysis results for the comparison "haptic expert demonstrations" vs. "no haptic expert demonstrations".

	k	N	g	p	Lower CI	Upper CI	Q
Haptic Demonstrations	7	142	0.27	=.41	-0.26	0.80	13.38, p<.05
Analysis by Outcome	7	142					
Force Application	2	22	0.90	=.35	-0.99	2.79	3.95, p<.05
Completion Time	2	52	0.17	=.08	-1.30	1.65	6.44, p<.05
Error Rate	3	68	0.09	=.71	-0.39	0.57	1.32, p=.52

The meta-analysis showed no significant effect of haptic expert demonstrations on overall performance ($g = 0.27$, $p = .51$), with moderately, though significantly, inconsistent results ($I^2 = 55.2\%$). An analysis by outcome variable showed no significant effects on force application ($g = 0.90$, $p = .35$), completion time ($g = 0.17$, $p = .08$), and error rate ($g = 0.09$, $p = .71$). Since there were only three or fewer studies in each category, the homogeneity analysis should be interpreted with caution.

5.4. Haptic assistance

A summary of the results of meta-analyses on the effects of haptic assistance on task performance is presented in Table 6 (see Appendix H. for further details).

Table 6. Summary of random-effects meta-analysis results for the comparison "haptic assistance present" vs. "haptic assistance absent".

	k	N	g	p	Lower CI	Upper CI	Q
Haptic Assistance	9	152	1.61	<.01	1.09	2.13	20.32, p<.01
Analysis by Outcome	14*	230					
Force Application	1	10	1.27	<.001	0.37	2.31	n/a
Completion Time	7	112	1.03	<.001	0.48	1.58	13.55, p<.05
Error Rate	6	108	1.55	<.001	0.99	2.10	11.93, p<.05

Note that several studies may be represented for more than one outcome variables in this analysis.

The meta-analysis indicates a large and statistically significant effect of haptic assistance functions on overall performance ($g = 1.61$, $p<.01$). Even though the homogeneity analysis implies significant heterogeneity in reported effect sizes ($I^2 = 60.6\%$), the 95% CI show that even the lower bound value is suggestive of a large effect. The analysis by outcome suggests that assistance functions are highly effective in the reduction of applied forces ($g = 1.276$, $p<.001$), task completion time ($g = 1.55$, $p<.001$) and error rate ($g = 1.55$, $p<.001$), with medium, albeit significant, inconsistency in ratings ($I^2_{time} = 55.7\%$; $I^2_{error} = 58.1\%$). However, it should be noted that only one study was available that investigated the effect of a haptic assistance function on force application (N=10).

6. Discussion

The four main applications of haptic signals, i.e. vibrotactile and force feedback from the remote or virtual environment, as well as haptic assistance and haptic expert demonstrations, were investigated in their effects on task performance in order to establish setup-independent effects and to compare the different applications in their relative effectiveness. The meta-analysis indicated that vibrotactile feedback was only effective in reducing task completion times, but neither forces nor errors were significantly reduced by it. Particularly noticeable is the large variance in the reported effect sizes of force measurements. Although a number of factors may cause this variance, including task dependence or differing methodology, the most likely explanation for this variance would seem to be that some vibrotactile devices may be less effective than other devices in reducing applied forces. This conclusion seems to be supported by empirical evidence discussed in Chapter II, Section 1.1, which suggests that if vibration frequency or location varies in order to convey information of intensity or direction, vibrotactile feedback may be less effective than a uniform signal that alerts the user of a required response.

Represented by the largest number of studies in the meta-analysis, force feedback from the remote environment appears to be not only more popular in the teleoperation community than vibrotactile feedback, but also more promising in terms of task performance improvement as it seems to exert a considerably larger effect on overall performance ($g_{\text{force feedback}} = 1.08$ vs. $g_{\text{vibrotactile}} = 0.75$). Although its effect on task completion time is slightly lower, force feedback shows demonstrably large effects on error rate and force reductions. Perhaps not surprisingly, force feedback appears to be particularly effective in reducing forces, suggesting that it is of some importance to simulate contact forces realistically in order to improve the users' ability to regulate precisely the forces that are applied to the remote or virtual environment.

Haptic assistance functions were found to constitute the most effective application of haptic signals in terms of task performance improvement. Not only displayed haptic assistance functions the largest overall effect on task performance, it further proved to be very effective in terms of each of the three investigated performance measures. As they were intended for in their design, haptic assistance functions showed the largest effect on error rate reduction. It was also found to be highly effective in terms of force reduction, although only one study investigated this aspect; hence, this finding may not be truly representative of the actual effect. The smallest effect size was reported for measures of timing, although this effect is still larger than are those reported in force feedback and vibrotactile studies.

Despite the overwhelming evidence testifying to the effectiveness of haptic assistance functions in improving task performance, one needs to take into consideration that all studies apart from one investigated some form of virtual fixture providing task-dependent assistance. Thus, little information is provided regarding task-facilitative effects of haptic assistance functions in unstructured environments. Since teleoperation systems are primarily used for tasks that require flexibility, further studies on task performance with task-independent assistance functions are urgently needed in order to validate their effectiveness.

In contrast to the consistent effectiveness of haptic assistance functions, haptic expert demonstrations were only consistent in showing no effect, as neither the overall performance, nor any of the individual performance measures were found to show significant effects. It should be noted, however, that the investigated sample was fairly small with only seven studies and only three or fewer effect sizes reported for each outcome variable. Nevertheless, considering that only the largest effect sizes were included from each of these studies, it can be said with some confidence that haptic expert demonstrations cannot be recommended for the improvement of user task performance in teleoperation and three-dimensional virtual reality systems as it does not lead to significant performance gains. It needs to be pointed out, however, that while haptic expert demonstrations were not found to be linked to superior work performance, measures of training success, for instance shorter training times, were not investigated in this context.

Although most studies focus on testing for statistical significance, it should once again be pointed out that effect sizes are equally important in the interpretation of findings. A statistically significant effect can still be unimportant in an applied context; it is only the effect size that informs of the practical significance of findings. Apart from allowing for direct comparisons of the impact of different variables on a particular measure, effect sizes also allow for simple cost-benefit analyses, if the standard deviation of the measure is known. Calculations differ, depending on the types of effect size and measures used. Transformation tables for effect sizes and examples for these calculations can be found in Lipsey & Wilson (2001, pp. 146-156).

In the interpretation of these findings, it should be stressed again, that this study was not intended to investigate average but maximum effects. Furthermore, the observed heterogeneity indicates that the reported pooled effect sizes may not represent their respective sample of effect sizes very well, since there is significantly more variability in these scores as could be explained by chance deviations in performance measures. As such, the present analyses provide little to no information about the improvement in performance that one could realistically expect from the implementation of haptic signals. Instead, the results reflect an optimistic estimate of the largest improvement that one could hope to achieve. As such, only the best-case scenario is described with the rationale that an investment into haptic interfaces cannot be recommended if the most optimistic results do not suggest an improvement in task performance. Nevertheless, ascertaining the average impact of haptic signals on task performance remains a priority of interface research; therefore, researchers are encouraged to report all results for this purpose, even if they are not statistically significant.

Noticeably high inconsistency was found in each analysis. One cannot discount the possibility that this heterogeneity in reported effect sizes reflects a substantial influence of a variable that was not considered in this study, for example, the technical specifications of the haptic interface or more general methodology issues. In part, this heterogeneity may also be attributed to the large number of studies included in some of these analyses as well as the fact that only maximum rather than mean effects were selected for inclusion. Although no explicit evidence was found for this assumption, it is also conceivable that older studies report fewer performance gains compared to studies with modern haptic interfaces. Hence, the reported effect sizes should be interpreted with caution, as they may not necessarily accurately represent the true nature of the effect of haptic signals on task performance in modern teleoperation and virtual systems. In any case, this heterogeneity in findings also reflects the subjective impression one gains when viewing the literature: that a wide disparity in methods, tasks and systems exists, which are difficult to unite under broader themes. Theoretical development is urgently needed to further the unification and integration of haptic interface research.

Finally, it needs to be pointed out that meta-analysis has been termed a Garbage-In Garbage-Out technique, meaning that the results are only reliable and meaningful to the extent that the included studies are of high quality. For the present analysis, selective criteria were chosen that aimed at ensuring high research quality of the included studies. However, in many cases, relevant information was simply missing. For example, the randomisation of experimental conditions or the planning of practice sessions with the system are rarely elaborated. Although the omission of some relevant information seems hardly avoidable considering the stringent space restrictions of most publications, researchers are encouraged to report all information that would be necessary for the calculation of effect sizes.

Despite its caveats, the analysis outlined in this chapter provide overwhelming evidence that haptic signals are capable of improving task performance, as it provided a quantitative summary of available research evidence and calculated its effectiveness over a broad range of settings, systems, task setups, researchers, and circumstances. It further made a direct comparison of the effectiveness of different haptic applications possible, and established the strengths and weaknesses of each of the four main applications of haptic signals investigated in this work. The results of this study are particularly robust and as such, they lend themselves well to the formulation of guidelines on the design and effective use of haptic signals in industrial applications.

7. Guidelines for the implementation and effective use of haptic signals

An empirical synthesis of the haptic interface evaluation literature indicated that the different haptic applications vary widely in their impact on task performance measures of timing, force application and error rate. Based on the meta-analysis of previous performance evaluations, a number of guidelines regarding the effective implementation of haptic interface applications were derived:

- In order to reduce task completion times, vibrotactile interfaces, which convey contact information from the remote environment, are recommended as a cost-effective alternative to kinaesthetic feedback devices.
- Force feedback from the remote environment is overall more effective than vibrotactile feedback. It is recommended for time-critical applications, but seems particularly suited to applications that require a precise regulation of forces on part of the user.
- For tasks which are particularly error prone or for which it is vital to reduce the risk of damaged material through uncoordinated movements, haptic assistance functions, for example virtual fixtures, are particularly recommended. While keeping performance errors to a minimum, they also speed up performance.
- Haptic expert demonstrations are not recommended as a measure of improving task performance; however, the literature suggests some benefits for training success.

Chapter V.
The HMI Design Guide

The methodological approach taken in this work allowed for the derivation of transparent guidelines with proven effectiveness. With the results of the investigations detailed in previous chapters, specific recommendations for the design and effective employment of haptic feedback devices could be made. These have been summarised and combined with already established guidelines for the design of the multi-modal human-machine interface in virtual and teleoperation systems. Building on previous efforts by Deml (2004), an interactive software package was implemented which aimed at providing specific and comprehensive advice to system developers regarding the design of efficient interfaces for their particular teleoperation system. The origin and development of the resulting HMI (Human-Machine Interface) Design Guide[15] is outlined in the following section.

1. Previous compilations of guidelines for the design of human-machine interfaces

Information that is relevant for the design of human-machine interfaces can be derived from a number of sources. For one, several data compendia have been compiled, which consist of condensed and categorised databases with information on human capabilities. The four-volume publication by Boff & Lincoln (1998) "Engineering Data Compendium: Human Perception and Performance" is one such example, which is recommended by a number of standard texts on human factors engineering, including Wickens, Lee, Liu, et al. (2004). In addition, numerous design standards have been proposed over the past decades, which specify design requirements as well as provide recommendations. An overview of some of the standards that are most relevant to human-machine interface design is provided in Table 7.

---

[15] For a copy of the HMI Design Guide, please contact the author (verena.nitsch@unibw.de).

Table 7. System design standards.

Standard	Year	Title
MIL-STD-1472 D	1989	Military Standard: human engineering, design criteria for military systems, equipment and facilities
NASA-STD-3000	1995	NASA Man-Systems Integration Standards
VDI /VDE 3850	2000	User-friendly design of useware for machines
NUREG 700	2002	Nuclear Regulatory Commission: Human-system interface design review guidelines
EN ISO 9241-110	2006	Ergonomic of human-system interaction - Part 110: Dialogue principles
EN ISO 9241-920	2009	Ergonomics of human-system interaction - Part 920: Guidance on tactile and haptic interactions
EN ISO 9241-210	2010	Ergonomics of human-system interaction – Part 210: Human-centred design for interactive systems

The considerable information presented in the data compendia and design standards provide comprehensive, yet very specific and oftentimes overly-detailed advice on the design of hard- and software. Consequently, the consultation of these compendia and standards oftentimes necessitate a rather time-consuming review of large quantities of irrelevant information in the search of recommendations that are potentially relevant to a particular system or context. Furthermore, generally no information on the origin and effect of the suggested measures is provided, so that it is often difficult to derive more abstract principles and guidelines that would apply to situations that are not specified in these texts. Difficulties in the design of multi-modal human-machine interfaces in teleoperation systems are further exacerbated, as specific standards for these systems do not yet exist. As a consequence of their limited applicability to the development of new teleoperation systems, existing standards tend to be largely disregarded. The idea was therefore conceived to compile recommendations that are specific to the design of human-machine interfaces in teleoperation systems and to implement a heuristic system, which will allow developers to quickly retrieve those guidelines that are most relevant to their particular system and application context. Once a general setup of the human-machine interface is decided upon, the developer may then follow up with specific recommendations detailed in the data compendia, design standards and research literature.

1.1. Multicriteria Assessment of Usability for Virtual Environments (MAUVE)

In recent years, a number of heuristic systems have been developed which aim at providing support for system developers specifically in the design and implementation of human-machine interfaces. Stanney, Mollaghasemi, Reeves, et al. (2003) developed a heuristic system for the assessment of the usability of virtual environments, MAUVE. Here, around 140 evaluative criteria derived mainly from the literature were used to predict the overall usability of a VR system based on the fit (or lack thereof) to these principles. Since some of these criteria were potentially in conflict, a relative weighting of particular usability criteria could also be made. The present work is only remotely related to MAUVE, as the latter was intended to be used by usability professionals rather than system developers. As such, it focuses on the evaluation of the consequences of a particular interface design, rather than providing specific advice on the design itself. Furthermore, since its focus is on virtual reality systems, it does not consider those aspects of design that are specific to teleoperation systems (e.g. time delay, the problem of excessive forces, etc.).

1.2. The Presence Design Guide

Inspired by MAUVE, Deml (2004) introduced the Presence Design Guide, which utilised a heuristic, computer-aided approach to the design of teleoperation and virtual systems. In principle, the Presence Design guide constitutes of two separate questionnaires. One questionnaire aims at measuring the teleoperation system user's sense of presence, i.e. the degree to which the user feels present in the remote/virtual environment. Originally developed by Scheuchenpflug (2001), this questionnaire is intended to be completed by the end user of the teleoperation/virtual system in question. Based on an automated evaluation of the respondent's answers, an overall percentage score that indicates the user's sense of presence with a particular system is provided and, based on the scores on three subscales of this questionnaire (i.e. interface quality, spatial perception, involvement), advice on possible measures to improve the user's sense of presence is given. The second questionnaire that is provided by the Presence Design Guide aims at providing support during the development phase. Based on an analysis of the various system- and task-criteria specifications provided by the system developer, the Presence Design Guide recommends specific in- and output devices and outlines important in- and output requirements. Most of the recommendations provided by the Presence Design Guide tend to focus on performance improvements and are supported by empirical literature.

Since Deml's work was presented, research on human performance with these systems has virtually exploded. In order to utilise the information gathered in this research, it was therefore necessary to synthesise the reliable research conducted in the past seven years and formulate new guidelines, as well as augment existing ones. Furthermore, due to technological limitations of haptic devices at the time, the Presence Design Guide only provided very limited advice on the use and design of haptic interfaces. Consequently, the HMI Design Guide has at its disposal a large number of guidelines that were not available at the time of the development of the Presence Design Guide. This added volume necessitated a complete restructuring of the functionality of the software, which allows for a more interactive, and in many cases, faster use compared to previous heuristic systems. Thus, although the HMI Design Guide draws on a comparably large collection of specific guidelines, those that are relevant for a particular system and application context can be quickly retrieved. Hence, although the HMI Design Guide incorporates many of the design principles featured in the Presence Design Guide by Deml (2004), the present work is not only a continuation of her work, but also imposes a new functional and structural design on the compiled guidelines. As such, it allows for the provision of specific and comprehensive advice on the design and effective employment of multi-modal human-machine interfaces in teleoperation systems.

2. Development of the HMI Design Guide

2.1. Collected guidelines

In line with the research outlined in this work, it was decided that the focus of the HMI Design Guide should be placed on human work performance with industrial teleoperation system. As such, only guidelines that were empirically proven to relate to the user's ability to accomplish a particular task goal were considered for inclusion. Guidelines, which refer to aspects of interface design that were not consistently linked to work performance, such as hard-/software aesthetics, joy of use and the user's sense of presence, were not considered. A number of guidelines were derived from studies that were conducted during the nine-year term (2002-2010) of the research project *Evaluation and Design of the Human-Machine Interface in Telepresence and Teleaction Systems*[16], which was initiated precisely for the purpose of deriving guidelines from experimental user studies. Further considered were guidelines proposed in narrative reviews of empirical literature (e.g. Hale & Stanney (2004)) as well as those derived from studies included in the meta-analysis described in Chapter IV and listed in the Reference Section, pp.150-166. Finally, the guidelines described in the *Presence Design Guide* by Deml (2004), and applicable guidelines detailed in the proceedings of the *Guidelines on Tactile and Haptic Interactions* (GOTHI) 2005 conference were considered for inclusion. In some cases, standard texts on human perception and cognition (Eysenck & Keane, 2000), and on human factors engineering (Wickens, Lee, Liu, & Gordon Becker, 2004; Salvendy (2006)) were also consulted.

2.2. Implementation

The guidelines were compiled in an Xml-database, for which a questionnaire-type Graphical User Interface (GUI) was devised using the Windows Forms Application of Visual C++ (.Net Framework 3.5). Each guideline stored in the database was queried with at least one question displayed by the interface.

2.3. Structure

The HMI Design Guide was designed to communicate with the user interactively via a questionnaire-based interface (see Figure 38).

---

[16] This research was part of a Collaborative Research Center project (SFB 453: High-Fidelity Telepresence and Teleaction) and funded by the German Research Foundation (DFG).

Figure 38. Screenshot of the Welcome Screen of the HMI Design Guide.

For the purpose of selecting those design recommendations that are most relevant to the developer's particular system, five key categories of questions were identified. These key categories are defined by the user based on the responses given to a number of multiple-choice items. The responses to most items are optional. If no answers are provided to these questions, the collected guidelines for this particular factor will not be displayed. Answered questions will activate or deactivate linked questions that are posed later in the questionnaire and select or deselect the guidelines associated with these questions. Thus, the questions that are posed vary, depending on the respondent's answers to previous questions. In effect, this means that the more information is provided by the respondent, the more guidelines will be selected and the more specific and relevant the design recommendations that form the output will be. The five key categories of the HMI Design Guide interface, that make this personalised approach to design possible, are specified in the following.

2.3.1. Work domain

Since some applications that are available in virtual systems are not practical for teleoperated systems, the respondent is first asked to specify, whether his or her application is intended to be used in virtual realities or a real environment. Depending on the answer, the respondent is then given the option of specifying the work domain of the application, choosing between surgery, (micro-/macro) assembly or on-orbit servicing as teleoperation applications, or, if the application is intended for virtual reality use, product development or the virtual simulation of a robotic teleoperation system (see Figure 39).

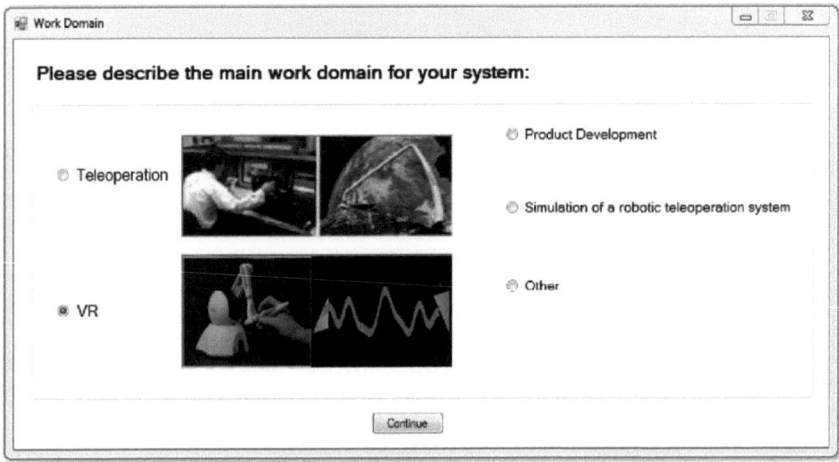

Figure 39. Screenshot example of the specification of work domain.

The selection of the work domain has consequences for later questions that are asked in the definition of contextual factors, as well as selecting directly guidelines for display. For example, if on-orbit servicing is chosen as intended field of application, the respondent will later be asked to specify the length of time-delay in the communication channel, while neglecting questions relating to gravity in the remote environment.

2.3.2. Prioritisation of work output requirements

In a next step, the respondent is asked to prioritise the work output requirements (see Figure 40): Is the emphasis of the task requirement on quantitative measures of work performance, such as task completion time, force application or handling error? Or is the emphasis placed on the end-users ability to form qualitative judgements, such as those required for texture or shape recognition, or discrimination ability?

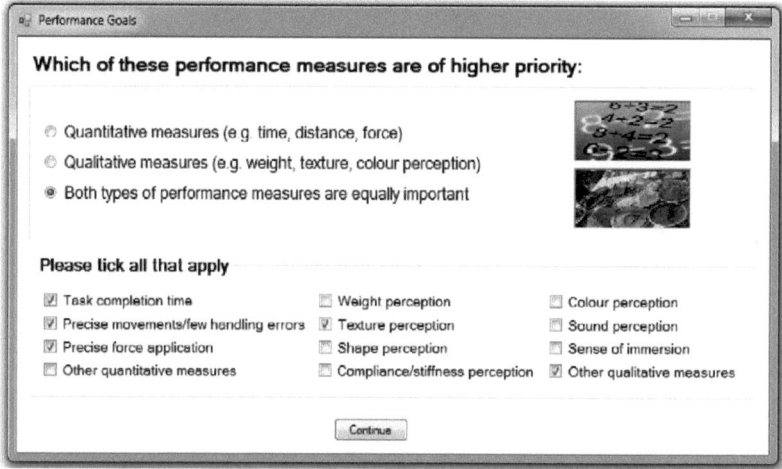

Figure 40. Screenshot example of the specification of performance goals.

An answer other than an equal weighting of priorities will lead to the deselection of irrelevant items. For example, if the emphasis is placed on movement speed, no further questions will be asked regarding the system's ability to display different textures or shapes.

2.3.3. Task analysis

Based on the specifications to previous questions, a detailed task analysis will be conducted, assessing the **G**oals, **O**perations, **M**ethods and **S**trategies (GOMS) of the task. The GOMS method has been in use for several decades in order to assess the efficiency of man-machine interactions in computer applications (Wickens, Lee, Liu, & Gordon Becker, 2004), and lends itself well as a framework for the analysis of task requirements and the methods used to fulfil these. The task analysis section is the most extensive of the HMI Design Guide, with up to 15 different specifications that can be provided here by the respondent. An example is shown in Figure 41.

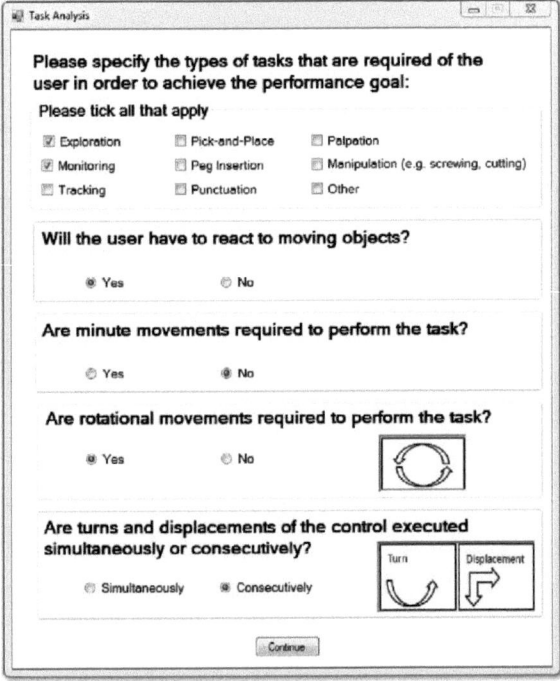

Figure 41. Screenshot example of the task analysis.

2.3.4. Specification of contextual factors

Based on the answers provided in the task analysis section, contextual factors relevant to the task are assessed, which include details of the remote environment, e.g. size and visibility of the workspace (see Figure 42). In addition, system specifications can be provided here, such as those pertaining to the sensors available in the remote environment, and time delay or bandwidth constraints pertaining to the communication channel between operator and teleoperator systems.

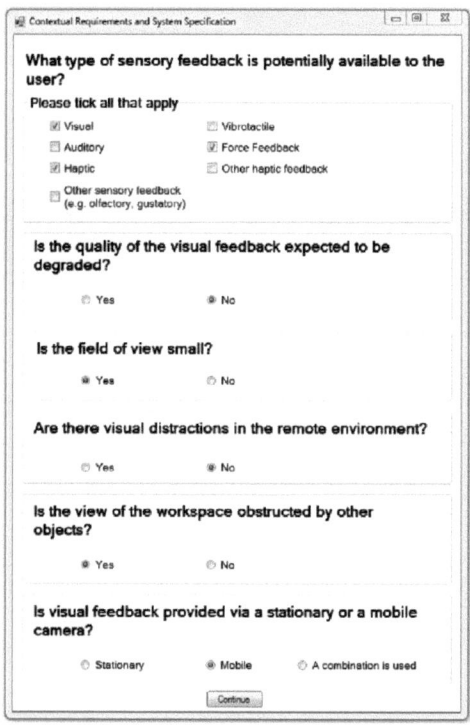

Figure 42. Screenshot example pertaining to the specification of contextual requirements and system specifications.

2.3.5. End-user specifications

Although, in general, personal factors were not found to be as important as system design and other contextual factors in determining the quality and efficiency of the human-machine interaction, some information may also be provided regarding the prospective end-user. Specifically, information regarding the proficiency level in handling the teleoperator and the user's dexterity and sensorimotor ability may be specified (Figure 43).

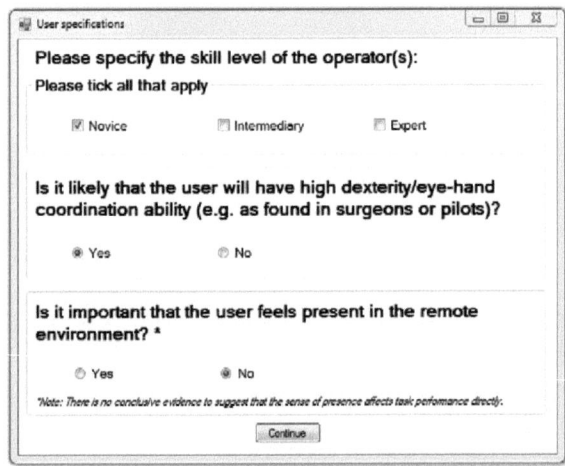

Figure 43. Screenshot example showing the specification of end-user characteristics.

These specifications are primarily used to advise on the application of particular feedback options, for example the use of haptic virtual fixtures for novice users or users with poor dexterity and eye-hand coordination.

2.3.6. Output

Based on the specifications provided by the developer, appropriate guidelines are retrieved from the database and listed according to the five input categories: A. Work domain, B. Performance goals, C. Task goals, operations, methods and strategies, D. Contextual and system requirements, and E. End-user specifications (see Figure 44).

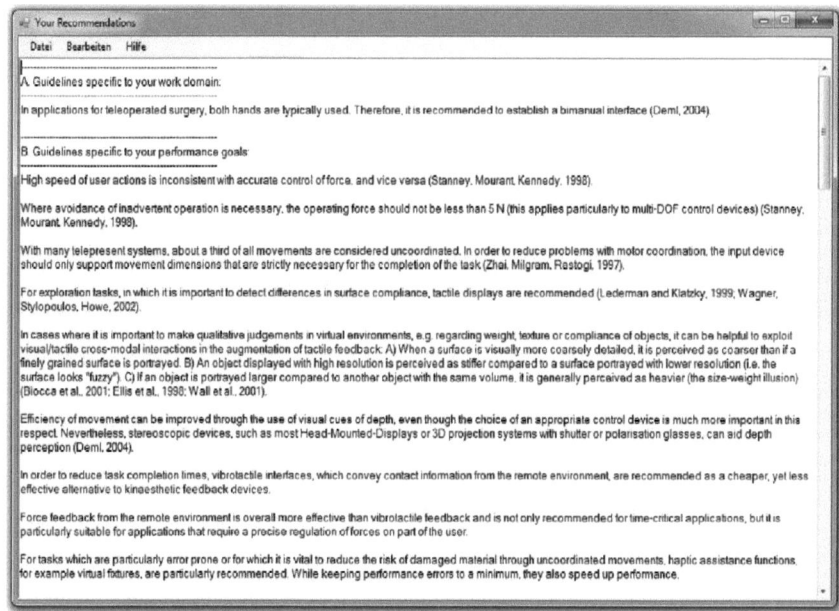

Figure 44. Screenshot example of the HMI Design Guide recommendations.

The displayed guidelines are automatically saved in form of a text document (.txt), so that they may be printed or edited later on. Alternatively, the developer has the option of displaying all guidelines that are available in the database. The user is also provided with the option of displaying the references to the included guidelines and is referred to existing standards and other relevant compilations of design guidelines.

2.4. An Example: Interface design for teleoperated minimally-invasive surgery according to the HMI Design Guide.

The specification of a teleoperated system for minimally-invasive surgery in this first step will select all guidelines that are relevant to this particular field of application. Next, priorities are assigned to the performance goals: do we need this system to perform quickly and accurately, or is it more important to provide information for qualitative judgments, like texture or weight discrimination? For this example, it is decided that quantitative and qualitative performance aspects are of equal importance. Based on these specifications, a detailed task analysis will then be conducted. In this example, the system is designed for the purpose of identifying clogged arteries through palpating of the tissue. It is specified that the task requires a lot of rotation, precise movements, and precise force control to palpate the tissue without damaging it. The end-user also needs to be able to differentiate between tissues of different compliances for this purpose. Based on this task analysis, contextual requirements of the remote work environment and the system are defined, for example, that the system needs to operate within a small workspace of poor visibility, and that visual feedback from the remote environment is displayed without noticeable time delay. Finally, some things about the generic end-user are specified, for example, that she or he would already be trained in the use of this system and that as surgeons, would likely possess superior dexterous ability.

Based on these few specifications, the design recommendations for the human-machine interface would include the use of an application-specific input device, which resembles as closely as possible to the target object, in this case an artery. An application-specific input device reduces mental effort required for the projection of intended movements, which is especially important if the task requires many rotation movements, but generally requires good dexterous abilities of the user. The control should be isotonic, since the workspace is small and a high degree of positional accuracy is necessary, and force feedback in three translational and three rotational degrees of freedom in the yaw, pitch and roll directions would be recommended to limit the maximum forces applied by the user to the tissue. Stereoscopic displays are recommended to improve depth perception in workspaces of poor visibility, and a visual point of reference should be provided for the user at all times, to help them orient themselves with the endoscopic camera. Finally, because the work with such systems requires a high amount of mental effort, no acoustic feedback should be used, so the user has to attend to fewer sensory channels simultaneously. Depending on further specifications, a number of more detailed recommendations, e.g. regarding the optimum operating force and sampling frequency, the use of visual cues of depth, and recommended luminance contrast.

3. Discussion

In an effort to synthesize past research efforts on human-machine interface design in a form that is accessible to system developers, a computer-aided, interactive design guide was implemented. The HMI Design Guide aims at providing comprehensive advice for the design of the human-machine interface, that is also relevant to the particular system which the teleoperation system developer is planning to use or construct. As such, it offers a structured approach for achieving effective human-machine interaction in teleoperation and virtual systems.

In the interpretation and application of the provided guidelines, several things should be kept in mind. For one, guidelines were based on experimental findings of effectiveness, however, apart from studies in which effect sizes were explicitly mentioned or could be calculated, no indication is given of the extent to which the measure are effective. It should also be taken into consideration that the majority of studies, which provide the basis for the guidelines, only tested individual components of the interface. As such, it is difficult to predict how different combinations of the tested components would affect human performance. For these reasons, it is vital that the design recommendations provided by this interactive guide are evaluated and validated in future research.

The presented guidelines can and should be considerably expanded upon, for example by incorporating other sensory modalities. In particular, the auditory modality is not yet sufficiently represented in human-machine interface guidelines of teleoperation systems as further evaluation studies investigating the effects of auditory signals on work performance are warranted. Gustatory and olfactory sensory modalities are also distinctly neglected in the literature on human-machine interaction, even though these senses also contribute to and shape the impression that we receive of our surroundings. Furthermore, although they seem to become increasingly relevant with a noticeable trend towards teleconferences and other multi-user applications, interfaces designed for multi-user systems were not explicitly considered, even though the majority of guidelines would seem to apply to cooperative systems as they do to single-user systems. System developers have also suggested that it would be valuable to provide a function that would allow them to assess the estimated efficiency of the human-machine interaction achieved with a particular control and display arrangement.

Chapter VI.
Summary and General Discussion

Teleoperation systems present numerous challenges to their large-scale employment in industrial applications. Work with these systems is oftentimes characterised by excessive force application and poor movement coordination, resulting in long performance times, high performance error rates and damage to material and equipment. In order to improve work performance with these systems, human-machine interfaces that effectively translate between human and machine in- and output signals need to be developed and appropriately employed. In particular, haptic interfaces promise to ameliorate observed performance problems by providing the user with tactile feedback from the remote environment and by supporting the user's movements directly with haptic signals to the operator devices. Overwhelmingly, system-centred approaches to the design of the haptic human-machine interface have been favoured so far, which emphasise technological affordances and challenges over practical utility. Hereby, developers have largely operated under the assumption, that it is sufficient to consider physical characteristics of human and machine sensors and effectors to ensure an effective translation process. However, this approach does not take into account cognitive processes that take place during human-machine interaction, i.e. the human interpretation of signals and in what way this interpretation informs the decision of the user to perform a particular behaviour. Since these cognitive processes are influenced by expectations (top-down processing) as much as they are determined by signal input (bottom-up processing), it seems that the application context is a critical factor in human-machine interaction, which needs to be considered in the design and effective employment of haptic interfaces.

Hence, in this work, an application-centred rather than system-centred approach is taken to the investigation of haptic human-machine interaction, with the aim of providing guidelines that advise system developers on the design and effective use of interfaces with haptic signal output capability in an industrial context. As such, technical specifications of the employed haptic devices were only marginally considered and the focus was placed instead on the effects of particular applications of haptic signals on behavioural output. Aiming to build on the large bulk of experimental work in this area, a methodological approach was chosen that was primarily empirically-deductive rather than theory-driven. In this context, four empirical investigations were conducted, with the purpose of furthering the understanding of haptic human-machine interaction, filling existing gaps in the literature as well as provide an empirical basis for the derivation of guidelines. Since these guidelines are particularly aimed at improving work performance with industrial teleoperation systems, particular care was taken to investigate the effects of haptic signals under realistic conditions. As such, all studies featured non-haptic control conditions with plausible setups, which could be realistically expected to exist in industrial teleoperation systems. In order to assess general, study-independent effects of these applications on task performance as well as ascertain the extent to which the investigated applications can improve various measures of work performance, the empirical studies were followed-up with meta-analytic investigations.

Two studies were conducted which investigated the effects of simulated haptic feedback from the remote environment on task performance measures. It was found that vibrotactile feedback would be particularly suited to tasks that require brief contact with objects, as it is the case in most pick-and-place tasks. In these situations, vibrotactile feedback was found to reduce the risk of damage to material and equipment in the remote environment that is caused by excessive contact forces. A meta-analytic investigation showed, however, that this effect might not generalise to other tasks, as other studies did not report significant effects on force regulation or, indeed, measures of performance accuracy. Speed, on the other hand, seems to increase with vibrotactile feedback for tasks that require frequent, yet intermittent contact with a surface, so that it may offer a cost-effective alternative to many other categories of haptic interfaces in these circumstances.

A more expensive category of haptic interfaces simulates not only tactile but also kinaesthetic feedback from the remote environment. This type of force feedback was found to be particularly suited to tasks that require constant contact with the remote environment, as it reduces task completion times and the application of excessive forces. As the meta-analysis showed, these effects were also found for other studies. It further indicated that force feedback can reduce performance errors due to improved coordination abilities. Although it was overall found to be more effective than vibrotactile feedback, force feedback is much more difficult to implement as it directly interferes with the kinematics of the control input.

Haptic signals cannot only be used to simulate feedback that is otherwise missing in teleoperation systems, but also to guide the user's movements. Despite their unproven effectiveness, studies are repeatedly implemented which use haptic expert demonstrations, whereby an expert's movements are pre-recorded and played back to novice users. While numerous studies attest to the effectiveness of haptic feedback from the remote environment during training, an empirical study in this work and a subsequent meta-analysis demonstrated rather convincingly that haptic expert demonstrations do not significantly improve motor coordination skills of novice users during subsequent task performance, nor do they seem to reduce training time. It seems that when it comes to the training of particular motor skills, active movements are more effective than passive movements and user control over movements should not be curtailed during training.

Haptic assistance functions are the most direct form of movement support. Hereby, haptic signals may simply be used to allow for a more comfortable control of teleoperator movements, for example by facilitating acceleration and deceleration movements, regardless of the movement requirements of the task goal. In contrast to this task-independent assistance, task-dependent assistance may be provided, for example in the form of virtual fixtures. This form of assistance only facilitates movements that are considered necessary in order to achieve the task goal, while inhibiting task-averse movements. Empirical evidence suggests that, in contrast to task-independent assistance is task-dependent assistance very effective in improving motor coordination, as it significantly reduces task completion times and movement errors. However, the study presented in this work also demonstrates how important it is to consider the extent to which the assistance curtails the user's autonomy over movements. If around 50% of movement control is left to the user, performance and perceived comfort improves, as does the flexibility of the system. A system which allows for flexible adaptation of the level of autonomy provided should be strived for.

Although haptic signals may be an effective measure in improving human coordination and force regulation ability of teleoperator movements, it is not yet clear which effect training has on this ability. The presented user studies certainly suggest that users become better with practice, regardless of the presence or absence of haptic signals. Previous studies, which investigated the problem of inefficient movement coordination, in particular, focused on intuitive coordination abilities, thus disregarding the effects of practice beyond those displayed by novice users after several trials. Hence, one might argue that many of the problems that were observed in the handling of teleoperation systems would diminish with sufficient practice, thus rendering current efforts to enhance the human-machine interaction through better interface design obsolete. Indeed, the remarkable adaptability of humans has lured system developers in the past into thinking that humans need to be made for machines rather than the other way around. This way of thinking, however, has been challenged by the increasing realisation that the physical design of a machine has an enormous impact on the human's ability to interpret machine output correctly and to react appropriately (Glendon, Clarke, & McKenna, 2006). In contrast, individual factors such as age, gender, and even sensorimotor coordination ability were found to contribute only marginally to one's ability to handle a particular system (Deml, 2004). Although human-machine interaction in teleoperation systems will most likely improve with practice to a certain point, the fact that even surgeons who have trained extensively with teleoperated surgical systems still report difficulties in the handling of these machines, suggests that better system design is needed to improve the efficiency of the human-machine interaction beyond that which can be achieved through practice alone.

Consequently, based on a synthesis of available research evidence, guidelines were devised which advise on the most efficient use of haptic signals in teleoperation systems, thus supporting system developers in the design of effective human-machine interfaces that are appropriate to their particular system. These were compiled and added to previously established guidelines on the design of multi-modal interfaces in form of the HMI Design guide, an interactive software package that allows system developers to employ human-machine interfaces that allow for the most efficient interaction between human and machine in teleoperation systems. Although the guidelines presented in this work were derived from empirical observations, further experimental studies need to be conducted so that these guidelines may be evaluated in their applicability to a wide range of systems.

The result presented in this work should not be interpreted without considering its limitations. For one, the quantitative measures of work performance, which were the focus of this work, are necessarily very restrictive in that they only provide a brief glimpse of haptic human-machine interaction in teleoperation systems. While measures of timing and force application are clearly defined, error rate is a very diverse measure, referring to position errors as well as more general failures in achieving a particular task goal. Although particular care was taken that the measures of error considered in the presented studies all reflect a general inability to precisely control the teleoperator's movements, it cannot be precluded that the chosen measures are less homogenous than assumed and that haptic signals may thus differ in their effects on particular measures of performance error. Hence, one should be aware that important performance benefits might not have been uncovered in the present work.

Secondly, an important measure of work performance has not been considered thus far: the ability of users to identify and discriminate objects based on qualitative assessments of objects, such as texture, shape, and compliance. In fact, performance problems that are caused by the inability to perform such assessments are widely reported; they are particularly prominent during telesurgery (Wagner, Howe, & Stylopoulos, 2002). These difficulties have largely been attributed to the absence of tactile feedback and as such, this aspect of work performance holds the most promise for haptic applications. The effectiveness of haptic signals in improving users' ability to make qualitative assessments of object properties is well-researched and well-documented, however. As such, they were not focus of the present investigation. Nevertheless, in considering the overall value of haptic signals for improving the human-machine interaction with teleoperation systems, the performance aspect of qualitative assessments should be taken into account as the primary beneficiary of haptic feedback.

Thirdly, the application-centred approach to the investigation of haptic human-machine interaction favoured in the present work was chosen to ensure that the behaviour observed in the featured studies is indicative of future work performance with these systems in an industrial context. This means that cognitive processes, which take place during the operation of industrial teleoperation systems and inform the user's behavioural decisions, were taken into consideration, but they were not explicitly investigated. As such, the present work can only provide tentative indications of thought processes that are likely to take place during the operation of teleoperation systems in industry. Furthermore, while the predictive validity of the presented studies is likely much higher than that of out-of-context laboratory studies, a number of factors which are likely to have a marked influence on real-life work performance with teleoperated systems, such as performance motivation and an organisation's safety culture, were not considered in these studies. Hence, as a next step, field studies need to be conducted which investigate the applicability of the proposed guidelines on actual work performance in industry and establish the financial costs and benefits of haptic human-machine interfaces.

Finally, related to the cognitive processes that take place during human-machine interaction are user perceptions of usefulness and acceptance of teleoperation systems, an area that has certainly been neglected thus far. The surveys that were incorporated in the various studies featured in the present work offer some tentative insight into these user perceptions; however, systematic investigations into qualitative assessments remain elusive. Yet, this aspect should not be neglected in the development of new technologies, as users' perceptions determine their decision to acquire and make use of these systems. Although one might argue that this is the domain of market research rather than system development, it cannot be ignored that the functional and physical design of a system has a marked influence on these decisions. Hence, user perceptions constitute a valuable source of input for the development of technological systems and as such warrant closer attention in the future.

In conclusion, haptic human-machine interfaces promise to enhance the effectiveness of the human-machine interaction in teleoperation system, and it appears that this promise is kept. The empirical studies outlined in the present work indicate that haptic signals can reduce the risk of damage through excessive force application and improve teleoperator movement coordination abilities - provided that they are applied appropriately. The application-centred approach that was taken in this work placed previous system-centred work into context and thus allowed for the derivation of transparent design guidelines with proven effectiveness in that context. It is hoped, that this work will be extended in the future, so that eventually a shift in focus towards user-centred design can take place. Further theoretical development on haptic human-machine interaction is also urgently required in order to provide a framework that guides future investigations of this topic. Only a methodological approach integrating theory and applied studies can unite the fragmented research efforts that currently prevail in the field of haptic human-machine interaction and thus effect improvement of work performance with industrial teleoperation systems.

Outlook

Although the engineering community appears to have taken a profound interest in human-machine interfaces that simulate haptic signals in recent years, most have not found their way into mainstream applications yet. Moreover, numerous applications of haptic signals have not been considered yet for systematic integration into the design of interfaces. For example, interfaces that stimulate body parts other than arms or hands, or those that utilise haptic icons for the transmission of information are still largely confined to research laboratories. Consequently, the full potential of haptic signal applications and their impact on future technologies cannot yet be fully anticipated.

Despite a recent surge in published material on haptic technology, and a long-standing tradition in the experimental research of physiological and cognitive aspects of the various human sensory modalities, it seems that research on applications of haptic interfaces and their effect on the human-machine interaction is still in its infancy. To further our understanding of haptic human-machine interaction and develop technology that builds on our accumulated knowledge and experience, increased interdisciplinary research is urgently needed. Although it becomes increasingly apparent that the development of future technological systems can only benefit from closer cooperation between the disciplines of psychology and engineering, the challenges of interdisciplinary work in this domain remain considerable.

As Wickens (2004, p. 32) had observed:

> "Poor design is common, and as our products become more technologically sophisticated, they frequently become more difficult to use. Even when designers attempt to consider human factors, they often complete the product design first and only then hand off the blueprint or prototype to a human factors expert. This expert is then placed in the unenviable position of having to come back with criticisms of a design that a person or design team has probably spent months and many thousands of dollars to develop. It is not hard to understand why engineers are less than thrilled to receive the results of a human factors analysis. They have invested in the design, clearly believe in the design, and are often reluctant to accept human factors recommendations. The process of bringing human factors analysis in at the end of the product design phase inherently places everyone involved at odds with one another. Because of the investment in the initial design and the designer's resistance to change, the result is often a product that is not particularly successful in supporting human performance, satisfaction, and safety."

Perhaps, transparent guidelines to human-machine interface design that are derived from empirical user studies, such as those established in this work, will encourage more system developers to consider principles of human-machine interaction during the design process, as well as support the efforts of human factors specialists in gaining a footing in the engineer's domain of system development.

References *

Aarno, D., Ekvall, S., & Kragic, D. (2005). Adaptive virtual fixtures for machine-assisted teleoperation tasks. *IEEE International Conference on Robotics and Automation*, (pp. 1151-1156).

Abbott, J. J., Hager, G. D., & Okamura, A. M. (2003). Steady-hand teleoperation with virtual fixtures. *12th IEEE International Workshop on Robot and Human Interactive Communication*, (pp. 145-151).

Abbott, J. J., Marayong, P., & Okamura, A. M. (2007). Haptic Virtual Fixtures for Robot-Assisted Manipulation. *Springer Tracts in Advanced Robotics, 28*, pp. 49-64.

Adams, R. J., & Hannaford, B. (2002). Control Law Design for Haptic Interfaces to Virtual Reality. *IEEE Transactions on Control Systems Technology, 10* (1), pp. 3-13.

* Adams, R. J., Klowden, D., & Hannaford, B. (2001). Virtual training for a manual assembly task. *Haptics-e*.

Akamatsu, M., & MacKenzie, I. S. (1996). Movement characteristics using a mouse with tactile and force feedback. *International Journal of Human-Computer Studies* (45), pp. 483-493.

Akamatsu, M., MacKenzie, I. S., & Hasbroucq, T. (1995). A comparison of tactile, auditory, and visual feedback in a pointing task using a mouse-type device. *Ergonomics* (38), pp. 816-827.

Albus, J. S., McCain, H. G., & Lurnia, R. (1989). NASA/NBS standard reference model for telerobot control system architecture (NASREM). In *NIST Technical Note 1235*. National Bureau of Standards.

Allison, L., & Jeka, J. J. (2004). Multisensory integration: resolving ambiguities for human postural control. In G. Calvert, C. Spence, & B. E. Stein (Eds.), *The Handbook of Multisensory Processes* (pp. 785-798). Cambridge, USA: MIT Press.

Almansa, A., Brenner, W., Wogerer, C., & Kiriakidis, G. (2004). From mechatronics to microsystems technology: European scale training in microhandling and -assembly. *IEEE International Conference on Mechatronics, ICM '04*, (pp. 242-246).

Aracil, R., Buss, M., Cobos, S., Ferre, M., Hirche, S., Kuschel, M., et al. (2007). The human role in telerobotics. In M. Ferre, M. Buss, R. Aracil, C. Melchiorri, & C. Balaguer (Eds.), *Advances in Telerobotics* (pp. 1-7). Berlin Heidelberg: Springer Verlag.

Aracil, R., Penin, L. F., Ferne, M., Jimenez, L. M., Barrientos, A., Santamaria, A., et al. (1995). ROBTET: A new teleoperated system for live-line maintenance. *7th International Conference on Transmission and Distribution Construction and Live-Line Maintenance*.

* Arsenault, R., & Ware, C. (2000). Eye-hand coordination with force feedback. *SIGCHI Conference on Human Factors in Computing Systems*, (pp. 408-414).

Bäckmann, L., & Nilsson, L. G. (1991). Effects of divided attention on free and cued recall of verbal events and action events. *Bulletin of the Psychonomic Society, 29* (1), pp. 51-54.

---

* denotes studies that were used in the meta-analysis (pp. 112 ff.)

Ballesteros, S. (2008). Implicit and explicit memory effects in haptic perception. In M. Grunwald (Ed.), *Human Haptic Perception: Basics and Applications* (pp. 183-197). Basel Boston Berlin: Birkhäuser Verlag.

Bau, O., Poupyrev, I., Israr, A., & Harrison, C. (2010). TeslaTouch: electrovibration for touch surfaces. *Proceedings of the 23rd Annual ACM Symposium on User Interface Software and Technology (UIST '10)*, (pp. 283-292).

Bertelson, P., & De Gelder, B. (2004). The psychology of multi-modal perception. In C. Spence, & J. Driver (Eds.), *Crossmodal space and crossmodal attention* (pp. 141-177). Oxford New York: Oxford University Press.

* Bethea, B. T., Okamura, A. M., Kitagawa, M., Fitton, T. P., Cattaneo, S. M., Gott, V. L., et al. (2004). Application of haptic feedback to robotic surgery. *Journal of Laparoendoscopic & Advanced Surgical Techniques, 14* (3), pp. 191-195.

* Bettini, A., Marayong, P., Lang, S., Okamura, A. M., & Hager, G. D. (2004). Vision-assisted control for manipulation using virtual fixtures. *IEEE Transactions on Robotics, 20* (6), pp. 953-966.

Biggs, J., & Srinivasan, M. (2002). Haptic Interfaces. In *Haptic Interfaces. Handbook of Virtual Environments* (pp. 93-116). London.

Birbaumer, N., & Schmidt, R. F. (2003). Bewegung und Handlung. In N. Birbaumer, & R. F. Schmidt, *Biologische Psychologie* (6 ed., pp. 255-295). Heidelberg: Springer Medizin Verlag.

Blackman, A. R. (1975). Test of additive factors method of choice reaction time analysis. *Perceptual and Motor Skills*, pp. 607-613.

Boff, K., & Lincoln, T. (1998). Engineering data compendium: Human perception and performance. Wright-Patterson Airforce Base, Ohio: Armstong Aerospace Medical Research Labaratory, AAMRL/NATO.

Book, W., & Love, L. (1999). Teleoperation, telerobotics, and telepresence. In S. Nof, *Handbook of Industrial Robotics* (Vol. 1, pp. 167-186). New York: John Wiley & Sons.

Borenstein, M., Hedges, L. V., Higgins, J. P., & Rothstein, H. R. (2009). *Introduction to Meta-analysis.* Chichester, UK: John Wiley & Sons, Ltd.

Bortz, J., & Döring, N. (2006). *Forschungsmethoden und Evaluation* (4 ed.). Heidelberg: Springer Medizin Verlag.

Bouguila, L., Ishii, M., & Sato, M. (2000). Effect of coupling haptics and stereopsis on depth perception in virtual environment. *First Workshop on Haptic Human Computer Interaction*, (pp. 54-62).

Brooks, F. P., Ouh-Young, M., & Batter, J. J. (1990). Project GROPE- Haptic displays for scientific visualisation. *Computer Graphics, 24* (4), pp. 177-185.

Brown, M. L., Newsome, S. L., & Glinert, E. P. (1989). An experiment into the use of auditory cues to reduce visual workload. *SIGCHI Conference on Human Factors in Computing Systems*, (pp. 339-346).

Bruno, N., & Cutting, J. E. (1988). Mini-modularity and the perception of layout. *Journal of Experimental Psychology: General* (117), pp. 161-170.

Burdea, G. C. (1996). *Force and Touch Feedback for Virtual Reality*. New York: John Wiley & Sons.

Burke, J. L., Prewett, M. S., Gray, A. A., Yang, L., Stilson, F. R., Coovert, M. D., et al. (2006). Comparing the effects of visual-auditory and visual-tactile feedback on user performance: a meta-analysis. *8th International Conference on Multimodal Interfaces*, (pp. 108-117).

Buss, M., Kuschel, M., Lee, K. K., Peer, A., Stanczyk, B., & Unterhinninghofen, U. (2006). High Fidelity Telepresence Systems: Design, Control, and Evaluation. *Joint International COE/HAM SFB-453 Workshop on Human Adaptive Mehcatronics and High-Fidelity Telepresence*.

* Cao, C. G., Zhou, M., Jones, D. B., & Schwaitzberg, S. D. (2007). Can surgeons think and operate with haptics at the same time? *Journal of Gastrointestinal Surgery, 11*, pp. 1564-1569.

* Chatterjee, A., Chaubey, P., Martin, J., & Thakor, N. (2008). Testing a prosthetic haptic feedback simulator with an interactive force matching task. *Journal of Prosthetics and Orthotics, 20* (2), pp. 27-34.

Chen, D. (2008, July/August). *Haptics for touch-enabled simulation and training*. Retrieved January 30th, 2011, from Patient Safey & Quality Healthcare: http://www.psqh.com/julaug08/haptics.html

* Cheng, L. T., Kazman, R., & Robinson, J. (1996). Vibrotactile feedback in delicate virtual reality operations. *Fourth ACM international Conference on Multimedia*, (pp. 243-251).

* Chmarra, M. K., Dankelman, J., Van den Dobbelsteen, J. J., & Jansen, F. W. (2008). Force feedback and basic laparoscopic skills. *Surgical endoscopy, 22* (10), pp. 2140-2148.

Chu, C.-C. P., Dani, T. H., & Gadh, R. (1997). Multi-sensory user interface for a virtual-reality-based computeraided design system. *Computer-Aided Design, 29* (10), pp. 709-725.

Cohen, J. (1988). *Statistical power analysis for the behavioural sciences* (2 ed.). New York: Academic Press.

Cohen, N. J., & Squire, L. R. (1980). Preserved learning and retention of pattern-analysing skill in amnesia: Dissociation of knowing how and knowing that. *Science, 210*, pp. 207-210.

Cohen, R. L. (1981). On the generality of some memory laws. *Scandinavian Journal of Psychology, 22*, pp. 267-281.

Cronbach, L. J. (1951). Coefficient alpha and the internal structure of the tests. *Psychometrika, 16* (3), pp. 297-334.

Debus, T., Jang, T.-J., Dupont, P., & Howe, R. (2002). Multi-channel vibrotactile display for teleoperated assembly. *IEEE International Conference on Robotics & Automation*, (pp. 592-597).

Deml, B. (2004). Telepräsenzsysteme: Gestaltung der Mensch-System-Schnittstelle. *Doctoral Thesis* . Neubiberg: Universität der Bundeswehr München.

* Deml, B., Ortmaier, T., & Seibold, U. (2005). The touch and feel in minimally invasive surgery. *IEEE International Workshop on Haptic Audio Visual Environments and their Applications*, (pp. 33-38).

* Dennerlein, J. T., & Yang, M. C. (2001). Haptic Force-Feedback Devices for the Office Computer: Performance and Musculoskeletal Loading Issues. *Human Factors, 43* (2), pp. 278-286.

Dennerlein, J. T., Howe, R. D., & Millman, P. A. (1997). Vibrotactile feedback for industrial telemanipulators. *ASME Dynamic Systems and Control Division, 61*, pp. 189-195.

Dennerlein, J. T., Shahion, E., & Howe, R. (2000). Vibrotactile feedback for an underwater teleoperated robot. *International Symposium on Robotics with Applications.*

Draper, J., Hemdon, J., & Moore, W. (1987). The implications of force reflection for teleoperation in space. *Conference on Space Applications of Artificial Intelligence and Robotics.*

Druyan, S. (1997). Effect of the kinesthetic conflict in promoting scientific reasoning. *Journal of Research in Science Teaching, 10*, pp. 1083-1099.

Duchaine, V., & Gosselin, C. (2007). General model of human-robot cooperation using a novel velocity based variable impedance control. *WorldHaptics*, (pp. 446-451).

* Edwards, G. W., Barfield, W., & Nussbaum, M. A. (2004). The use of force feedback and auditory cues for performance of an assembly task in an immersive virtual environment. *Virtual Reality, 7* (2), pp. 112-119.

Engelkamp, J., & Krumnacker, H. (1980). Imaginale und motorische Prozesse beim Behalten verbalen Materials. *Zeitschrift für experimentelle und angewandte Psychologie, 12*, pp. 511-533.

Engelkamp, J., & Zimmer, H. D. (1985). Motor programs and their relation to semantic memory. *German Journal of Psychology, 9*, pp. 239-254.

Ernst, M. O., & Banks, M. S. (2002). Humans integrate visual and haptic information in a statistically optimal fashion. *Nature* (415), pp. 429-433.

Eysenck, M. W., & Keane, M. T. (2000). *Cognitive Psychology* (4 ed.). New York: Taylor & Francis Psychology Press.

* Fattouh, A., Sahnoun, M., & Bourhis, G. (2004). Force feedback joystick control of a powered wheelchair: preliminary study. *IEEE International Conference on Systems, Man and Cybernetics*, (pp. 2640-2645).

* Feller, R. L., Lau, C. K., Wagner, C. R., Perrin, D. P., & Howe, R. D. (2004). The effect of force feedback on remote palpation. *IEEE International Conference on Robotics and Automation*, (pp. 782-788).

* Feygin, D., Keehner, M., & Tendick, F. (2002). Haptic guidance: Experimental evaluation of a haptic training method for a perceptual motor skill. *10th Symposium On Haptic Interfaces For Virtual Environments & Teleoperator Systems (HAPTICS).*

Field, A. (2009). *Discovering statistics using SPSS* (3 ed.). London, UK: Sage.

Fitts, P. M., & Deininger, R. L. (1954). S-R Compatibility: Correspondence among paired elements within stimulus and response codes. *Journal of Experimental Psychology, 48* (6), pp. 483-492.

Fitts, P. M., & Posner, M. (1967). *Human Performance.* Belmont, CA: Brooks/Cole.

Fitts, P. M., & Seeger, C. M. (1953). S-R compatibility: Spatial characteristics of stimulus and response codes. *Journal of Experimental Psychology, 46* (3), pp. 199-210.

* Forsberg, L. (2008). Increasing performance and reducing the visual information overload when using a computer mouse with the help of vibrotactile feedback. *Umea's 12th Student Conference in Computing Science*, (pp. 101-109).

* Forsyth, B. A., & MacLean, K. E. (2006). Predictive haptic guidance: intelligent user assistance for the control of dynamic tasks. *IEEE Transactions on Visualization and Computer Graphics, 12* (1), pp. 103-113.

Gallagher, A. G., & Cates, C. U. (2004). Approval of virtual reality training for carotid stenting: what this means for procedural-based medicine. *Journal of the American Medical Association, 292*, pp. 3024-3026.

Gehring, W. J., Gratton, G., Coles, M. G., & Donching, E. (1992). Probability effects on stimulus evaluation and response processes. *Journal of Experimental Psychology: Human Perception and Performance, 18* (1), pp. 198-216.

Geiger, L., Popp, M., Färber, B., Artigas, J., & Kremer, P. (2010). The influence of telemanipulation-systems on fine motor performance. *Third International Conference on Advances in Computer- Human Interactions*, (pp. 44-49).

Gentaz, E., & Hatwell, Y. (2008). Haptic perceptual illusions. In M. Grunwald (Ed.), *Human Haptic Perception: Basics and Applications* (pp. 223-233). Basel Boston Berlin: Birkhäuser Verlag.

* Gerovich, O., Marayong, P., & Okamura, A. M. (2004). The effect of visual and haptic feedback on computer-assisted needle insertion. *Computer Aided Surgery, 9* (6), pp. 243-249.

* Gibo, T. L., Verner, L. N., Yuh, D. D., & Okamura, A. M. (2009). Design considerations and human-machine performance of moving virtual fixtures. *IEEE International Conference on Robotics and Automation*, (pp. 671-676).

Gibson, J. J. (1966). *The senses considered as perceptual systems.* Boston: Houghton Mifflin.

Gill, S. A., & Ruddle, R. A. (1998). Using virtual humans to solve real ergonomic design problems. *IEEE International Conference on Simulation*, (pp. 223-229).

Gillespie, R., O'Modhrain, M., Tang, P., Zaretzky, D., & Pham, C. (1997). The virtual teacher. *Conference on ASME Dynamic Systems and Control Division*, (pp. 171-178).

Glendon, A. I., Clarke, S. G., & McKenna, E. F. (2006). *Human safety and risk management* (2 ed.). London New York: Taylor & Francis Group.

Goertz, R., & Thompson, R. (1954). Electronically controlled manipulator. *Nucleonics, 12* (11), pp. 46-47.

Goodwin, A. W., & Wheat, H. E. (2008). Physiological mechanisms of the receptor system. In M. Grunwald (Ed.), *Human Haptic Perception: Basics and Applications* (pp. 93-102). Basel Boston Berlin: Birkhäuser Verlag.

Grantcharov, T. P., Kristiansen, V. B., Bendix, J., Bardram, L., Rosenberg, J., & Funch-Jensen, P. (2004). Randomised clinical trial of virtual reality simulation for laparoscopic skills training. *British Journal of Surgery, 91* (2), pp. 146-150.

Greenhouse, S. W., & Geisser, S. (1959). On methods in the analysis of profile data. *Psychometrika* (24), pp. 95-112.

Guidelines on tactile and haptic interactions. (2005). Retrieved October 28th, 2008, from Proceedings of GOTHI 05: http://userlab.usask.ca/GOTHI/GOTHI-05 Proceedings.html

Gunn, C., Hutchins, M., Stevenson, D., Adcock, M., & Youngblood, P. (2005). Using collaborative haptics in remote surgical training. *First Joint Eurohaptics Conference and Sysmposium on Haptic Interfaces for Virtual Environment and Teleoperator Systems.*

* Gupta, R., Sheridan, T. B., & Whitney, D. (1997). Experiments using multimodal virtual environments in design for assembly analysis. *Presence: Teleoperators and Virtual Environments, 6* (3), pp. 318-338.

Halata, Z., & Baumann, K. I. (2008). Anatomy of receptors. In M. Grunwald (Ed.), *Human Haptic Perception: Basics and Applications* (pp. 85-92). Basel Boston Berlin: Birkhäuser Verlag.

Hale, K. S., & Stanney, K. M. (2004). Deriving haptic design guidelines from human physiological, psychophysical, and neurological foundations. *IEEE Computer Graphics and Applications, March/April*, pp. 33-39.

Hasser, C. J., Goldenberg, A. S., Martin, K. M., & Rosenberg, L. B. (1998). User performing a GUI pointing task with a low-cost force-feedback computer mouse. *Conference on ASME Dynamics and Control Division*, (pp. 151-156).

Hedges, L. V., & Olkin, I. O. (1985). *Statistical methods for meta-analysis.* San Diegao, CA: Academic Press, Inc.

Hedicke, V. (2000). Multimodalität in Mensch-Maschine-Schnittstellen. In K. P. Timpe, T. Jürgensohn, & H. Kolrep (Eds.), *Mensch-Maschine-Systemtechnik* (pp. 203-232). Düsseldorf: Symposium Publishing.

* Heijnsdijk, E. A., Pasdeloup, A., Van der Pijl, A. J., Dankelman, J., & Gouma, D. J. (2004). The influence of force feedback and visual feedback in grasping tissue laparoscopically. *Surgical Endoscopy, 18* (6), pp. 980-985.

Hendry, S. H., Hsiao, S. S., & Bushnell, M. C. (1999). Somatic Sensation. In M. J. Zigmond, F. E. Boom, S. C. Landis, J. L. Roberts, & L. R. Squire (Eds.), *Fundamental Neuroscience* (pp. 761-787). London San Diego: Academic Press.

Higgins, J. P., & Thompson, S. G. (2002). Quantifying heterogeneity in a meta-analysis. *Statistics in Medicine, 21*, pp. 1539-1558.

Hirzinger, G., Brunner, B. K., Landzettel, K., Sporer, N., Butterfaß, J., & Schedl, M. (2003). Space robotics- DLR's telerobotic concepts, lightweight arms and articulated hands. In *Autonomous Robots* (Vol. 14, pp. 127-145). Kluwer Academic Publishers.

Hogg, M., & Vaughan, G. (2002). *Social Psychology* (3 ed.). UK: Prentice Hall.

Howitt, D., & Cramer, D. (2003). *An introduction to statistics in psychology* (2 ed.). Harlow, England: Prentice Hall.

* Hurmuzlu, Y., Ephanov, A., & Stoianovici, D. (1998). Effect of a pneumatically driven haptic interface on the perceptional capabilities of human operators. *Presence: Teleoperators and Virtual Environments, 7* (3), pp. 290-307.

Huynh, H., & Feldt, L. S. (1976). Estimation of the Box correction for degrees of freedom from sample data in randomsied block and split-plot designs. *Journal of Educational Statistics, 1* (1), pp. 69-82.

Ishii, M., Nakata, M., & Sato, M. (1994). Networked SPIDAR: A networked virtual environment with visual, auditory, and haptic interactions. *Presence: Teleoperators and Virtual Environments, 3* (4), pp. 351-359.

Itoh, T., Kosuge, K., & Fukuda, T. (1995). Human-machine cooperative telemanipulation with motion and force scaling using task-oriented virtual tool dynamics. *IEEE Transactions on Robotics and Automation, 16* (5), pp. 505-516.

Iwata, H. (2008). History of haptic interface. In M. Grunwald, *Human Haptic Perception* (pp. 355-361). Basel, Switzerland: Birkhäuser Verlag.

Iwata, H., Yano, H., Nakaizumi, F., & Kawamura, R. (2001). Project FEELEX: adding haptic surface to graphics. *28th Annual Conference on Computer Graphics and Interactive Techniques*, (pp. 469-475).

Iwata, T. (2001). Recent Japanese activities in space automation & robotics- An overview. *6th International Symposium on Artificial Intelligence and Robotics & Automation in Space.*

Jacoff, A., Messina, E., Weiss, B. A., Tadokoro, S., & Nakagawa, Y. (2003). Test arenas and performance metrics for urban search and rescue robots. *Intelligent Robots and Systems, 3*, pp. 3396-3403.

Jansson, G., & Öström, M. (2004). The effects of co-location of visual and haptic space on judgments of form. *EuroHaptics 2004*, (pp. 516-519).

Jentsch, F., & Bowers, C. (1998). Evidence for the validity of PC-based simulations in studying aircrew coordination. *International Journal of Aviation Psychology* (8), pp. 243-260.

Jokela, T., Iivari, N., Matero, J., & Karukka, M. (2003). The standard of user-centered design and the standard definition of usability: analyzing ISO 13407 against ISO 9241-11. *Proceedings of the Latin American Conference on Human-Computer Interaction*, (pp. 53-60).

Joly, L., & Andriot, C. (1995). Imposing motion constraints to a force reflecting telerobot through real-time simulation of a virtual mechanism. *IEEE International Conference on Robotics and Automation*, (pp. 357-362).

Jones, L. A., & Sarter, N. B. (2008). Tactile displays: guidance for their design and application. *Human Factoros: The Journal of the Human Factors and Ergonomics Society, 50* (9), pp. 90-111.

Jütte, R. (2008). Haptic perception: an historical approach. In M. Grunwald (Ed.), *Human Haptic Perception: Basics and Applications* (pp. 3-13). Basel Boston Berlin: Birkhäuser Verlag.

Kawamura, K., & Iskarous, M. (1994). Trends in service robots for the disabled and the elderly. *IEEE/RSJ/GI International Conference on Intelligent Robots and Systems*, (pp. 1647-1654).

Kazi, A. (2001). Operator Performance in Surgical Telemanipulation. *Presence: Teleoperators and Virtual Environments, 10* (5), pp. 495-510.

Kern, T. A. (2009). *Engineering Haptic Devices.* (T. A. Kern, Ed.) Heidelberg London New York: Springer.

Kim, T., & Hinds, P. (2006). Who should I blame? Effects of autonomy and transparency on attributions in human-robot interaction. *15th IEEE International Symposium on Robot and Human Interactive Communication*, (pp. 80-85).

Kim, W. S. (1992). A new scheme of force reflecting control. *Fifth Annual Workshop on Space Operations Applications and Research, 1*, pp. 254-261.

Kitagawa, M., Dokko, D., Okamura, A. M., & Yuh, D. D. (2005). Effect of sensory substitution on suture-manipulation forces for robotic surgical systems. *Journal of Thoracic and Cardiovascular Surgery, 129* (1), pp. 151-158.

Kitagawa, M., Okamura, A. M., Bethea, B. T., Gott, V. L., & Baumgartner, W. A. (2002). Analysis of suture manipulation forces for teleoperation with force feedback. *Medical Image Computing and Computer-Assisted Intervention, 2488*, pp. 155-162.

KIVA Systems. (2011). Retrieved November 21st, 2011, from http://www.kivasystems.com

Klatzky, R. L., & Lederman, S. J. (1999). Tactile roughness perception with a rigid link interposed between skin and surface. *Perception & Psychophysics, 61* (4), pp. 591-607.

Klatzky, R. L., & Lederman, S. J. (2003). Touch. In A. Healy, A. Proctor, & I. Weiner (Ed.), *Handbook of Psychology* (Vol. 4, pp. 147-176). West Sussex: John Wiley & Sons, Inc.

Klatzky, R. L., Lederman, S. J., & Metzger, V. A. (1985). Identifying objects by touch: an "expert system". *Perception & Psychophysics, 37* (4), pp. 299-302.

Klatzky, R. L., Lederman, S. J., & Reed, C. (1987). There's more to touch than meets the eye: the salience of object attributes for haptics with and without vision. *Journal of Experimental Psychology, 116* (4), pp. 356-369.

Kline, P. (1999). *The Handbook of Psychological Testing* (2 ed.). London: Routledge.

Kontarinis, D. A., & Howe, R. D. (1995). Tactile display of vibratory information in teleoperation and virtual environments. *Presence: Teleoperators and Virtual Environments, 4* (4), pp. 387-402.

Kornblum, S., Hasbroucq, T., & Osman, A. (1990). Dimensional overlap: Cognitive basis for stimulus-response compatibility- A model and taxonomy. *Psychological Review, 97* (2), pp. 253-270.

Kujala, S. (2003). User involvement: A review of the benefits and challenges. *Behaviour and Information Technology, 22* (1), pp. 1-16.

Laird, R. T., Bruch, M. H., West, M. B., Ciccimaro, D. A., & Everett, H. R. (2000). *Issues in vehicle teleoperation for tunnel and sewer reconnaissance.* Ft. Belvoir Defense Technical Information Center APR.

Lathan, C. E., & Tracey, M. (2002). The effects of operator spatial perception and sensory feedback on human-robot teleoperation performance. *Presence: Teleoperators and Virtual Environments, 11* (4), pp. 368-377.

Laycock, S. D., & Day, A. M. (2003). Recent developments and applications of haptic devices. *Computer Graphics Forum, 22* (2), pp. 117-132.

Lederman, S., & Klatzky, R. (2009). Haptic perception: A tutorial. *Attention, Perception & Psychophysics, 71* (7), pp. 1439-1459.

* Lee, S., Sukhatme, G. S., Kim, G. J., & Park, C. M. (2002). Haptic control of a mobile robot: A user study. *IEEE/RSJ International Conference on Intelligent Robots and Systems*, (pp. 2867-2874).

Lefcourt, H. M. (1973). The function of the illusions of control and freedom. *American Psychologist, 28* (5), pp. 417-425.

Letier, P., Avraam, M., Veillerette, S., Horodinca, M., DeBartolomei, M., Schiele, A., et al. (2008). SAM: A 7-DOF portable arm exoskeleton with local joint control. *IEEE/RSJ International Conference on Intelligent Robots and Systems*, (pp. 3501-3506).

Lewkowicz, D. J., & Kraebel, K. S. (2004). The value of multisensory redundancy in the development of intersensory perception. In G. Calvert, C. Spence, & B. E. Stein (Eds.), *The Handbook of Multisensory Processes* (pp. 655-678). Cambridge USA: MIT Press.

Lin, H. C., Mills, K., Kazanzides, P., Hager, G. D., Marayong, P., Okamura, A. M., et al. (2006). Portability and applicability of virtual fixtures across medical and manufacturing tasks. *IEEE International Conference on Robotics and Automation*, (pp. 225-230).

* Lindeman, R. W., Sibert, J. L., & Hahn, J. K. (1999). Towards usable VR: an empirical study of user interfaces for immersive virtual environments. *SIGCHI Conference on Human Factors in Computing Systems*, (pp. 64-71).

Linvill, J. G., & Bliss, J. C. (1966). A direct translation reading aid for the blind. *Proceedings of the Intstitute of Electrical and Electronics Engineers, 54*, pp. 40-51.

Lipsey, M. W., & Wilson, D. B. (2001). *Practical meta-analysis.* London New Delhi: Sage Publications, Inc.

Lipsey, M. W., & Wilson, D. B. (1993). The efficacy of psychological, educational, and behavioural treatment: Confirmation from meta-analysis. *American Psychologist* (48), pp. 1181-1209.

* Liu, J., Cramer, S. C., & Reinkensmeyer, D. J. (2006). Learning to perform a new movement with robotic assistance: comparison of haptic guidance and visual demonstration. *Journal of NeuroEngineering and Rehabilitation, 3* (20), pp. 1-10.

Liu, Y. C. (2001). Comparative study of the effects of auditory, visual and multimodality displays on drivers' performance in advanced traveller information systems. *Ergonomics, 44* (4), pp. 425-442.

Loomis, J., & Lederman, S. J. (1986). Tactual perception. In K. Boff, L. Kaufman, & J. Thomas (Eds.), *Handbook of Perception and Human Performance.* New York: Wiley.

MacLean, K. E. (2000). Designing with haptic feedback. *IEEE International Conference on Robotics and Automation*, (pp. 783-788).

Mao, J.-Y., Vredenburg, K., Smith, P. W., & Carey, T. (2005). The state of user-centered design practice. *Communications of the ACM, 48* (3), pp. 105-109.

Marayong, P., & Okamura, A. M. (2004). Speed-accuracy characteristics of human-machine cooperative manipulation using virtual fixtures with variable admittance. *Human Factors, 46* (3), pp. 518-532.

Martin, J., & Savall, J. (2005). Mechanisms for haptic torque feedback. *First Joint Eurohaptics Conference and Symposium on Haptic Interfaces for Virtual Environment and Teleoperator Systems*, (pp. 611-614).

Massimino, M. J., & Sheridan, T. B. (1993). Sensory substitution for force feedback in teleoperation. *Presence: Teleoperators and Virtual Environments, 2* (4), pp. 344-352.

Massimino, M. J., & Sheridan, T. B. (1992). Sensory substitution of force feedback for the human-machine interface in space teleoperation. *43rd Congress of the International Astronautical Federation, Paper IAF/iAA-92--0246.*

Massimino, M. J., Sheridan, T. B., & Roseborough, J. B. (1989). One hand tracking in six degrees of freedom. *IEEE International Conference on Systems, Man and Cybernetics*, (pp. 498-503).

Mattos, L., & Caldwell, D. (2009). Interface design for Microbiomanipulation and teleoperation. *2nd International Conference on Advances in Computer-Human Interactions*, (pp. 342-347).

Mauter, G., & Katzki, S. (2003). The application of operational haptics in automotive engineering. *Business Briefing: Global Automotive Manufacturing & Technology*, pp. 78-80.

* Mayer, H., Nagy, I., Knoll, A., Braun, E. A., Bauernschmitt, R., & Lange, R. (2007). Haptic feedback in a telepresence system for endoscopic heart surgery. *Presence: Teleoperators and Virtual Environments, 16* (5), pp. 459-470.

McCain, H. G., Andary, J. F., Hewitt, D. R., & Haley, D. C. (1991). The space station freedom flight telerobotic servicer: The design and evolution of a dexterous space robot. *Acta Astronautica, 24*, pp. 45-54.

Micaelli, A., Bidard, C., & Andriot, C. (1998). Decoupling control based on virtual mechanisms for telemanipulation. *IEEE International Conference on Robotics and Automation*, (pp. 1924-1931).

Mikropoulos, T. A., & Nikolou, E. (1996). A virtual hand with tactile feedback for virtual learning environments. *World Conference on Educational Multimedia and Cybermedia, 792.*

Moore, C. A., Peshkin, M. A., & Colgate, J. E. (2003). Cobot implementation of virtual paths and 3-D virtual surfaces. *IEEE Transactions on Robotics and Automation, 19* (2), pp. 347-351.

* Morris, D., Tan, H., Barbagli, F., Chang, T., & Salisbury, K. (2007). Haptic feedback enhances force skill learning. *Second Joint EuroHaptics Conference and Symposium on Haptic Interfaces for Virtual Environment and Teleoperator Systems.*

Muthig, K. P. (1990). Informationsaufnahme und Informationsverarbeitung. In C. G. Hoyos, & B. Zimolong (Eds.), *Ingenieurspsychologie* (pp. 92-114). Göttingen: Hogrefe.

Nakamura, H., & Honda, T. (2006). Power assist system for a industrial use. *Journal of the Society of Instrument and Control Engineering, 45* (5), pp. 445-448.

NASA-JPL. (2004). 21st century complete guide to robotics. (P. Management, Ed.)

Nielsen, J. (1993). *Usability Engineering.* San Diego, London, San Franciso: Academic Press.

* Nitsch, V., Passenberg, C., Peer, A., Buss, M., & Färber, B. (2010). Assistance functions for collaborative haptic interaction in virtual environments and their effect on performance and user comfort. *1st International Conference on Applied Bionics and Biomechanics.*

* Oakley, I., McGee, M. R., Brewster, S., & Gray, P. (2000). Putting the feel in look and feel. *SIGCHI Conference on Human Factors in Computing Systems*, (pp. 415-422).

Okamura, A. M., Richard, C., & Cutkosky, M. R. (2002). Feeling is believing: using a force feedback joystick to teach dynamic systems. *Journal of Engineering Education*, pp. 345-349.

O'Malley, M. K., & Ambrose, R. O. (2003). Haptic feedback applications for Robonaut. *Industrial Robot: An International Journal, 30* (6), pp. 531-542.

* O'Malley, M. K., Gupta, A., Gen, M., & Li, Y. (2006). Shared control in haptic systems for performance enhancement and training. *Journal of Dynamic Systems, Measurement, and Control, 128* (1), pp. 75-85.

Palmer, C., & Meyer, R. K. (2000). Conceptual and motor learning in music performance. *Psychological Science, 11* (1), pp. 63-68.

Park, S. S., Howe, R. D., & Torchiana, D. F. (2001). Virtual fixtures for robot-assisted minimally-invasive cardiac surgery. *Fourth International Conference on Medical Image Computing and Computer-Assisted Intervention*, (pp. 14-17).

Passenberg, C., Peer, A., & Buss, M. (2010). A survey of environment-, operator-, and task-adapted controllers for teleoperation systems. *Journal of Mechatronics, 20* (7), pp. 787-801.

Payandeh, S., & Stanisic, Z. (2002). On application of virtual fixtures as an aid for telemanipulation and training. *10th Symposium on Haptic Interfaces for Virtual Environments and Teleoperator Systems*, (pp. 18-23).

* Payette, J. (1996). Evaluation of a force feedback (haptic) computer pointing device in zero gravity. *ASME Dynamics Systems and Contol Division* (58), pp. 547-553.

Peer, A., & Buss, M. (2008). A new admittance type haptic interface for bimanual manipulation. *IEEE/ASME Transactions on mechatronics, 2008, 13* (4), pp. 416-428.

Perreault, J., & Cao, C. G. (2006). Effects of vision and friction in laparoscopic surgery. *Human Factors, 48* (3), pp. 574-586.

* Petzold, B., Zaeh, M. F., Faerber, B., Deml, B., Egermeier, H., Schilp, J., et al. (2004). A study on visual, auditory, and haptic feedback for assembly tasks. *Presence: Teleoperators and virtual environments, 13* (1), pp. 16-21.

Pezzementi, Z., Okamura, A., & Hager, G. (2007). Dynamic guidance with pseudoadmittance virtual fixtures. *IEEE International Conference on Robotics and Automation*, (pp. 1761-1767).

Pfister, H., & Laws, P. (1995). Kinesthesia-1: Apparatus to Experience in 1-D Motion. *Physics Teacher, 33* (4), pp. 214-220.

Piedboeuf, J. C., & Dupuis, E. (2001). Recent Canadian activities in space automation & robotics - an overview. *7th International Symposium on Artificial Intelligence and Robotics and Automation in Space.*

Pongrac, H. (2008). Gestaltung und Evaluation von virtuellen und Telepräsenzsystemen an Hand von Aufgabenleistung und Präsenzempfinden. *Doctoral Thesis.* Neubiberg: Universität der Bundeswehr München.

* Radi, M., Reiter, A., Zaidan, S., Nitsch, V., Faerber, B., & Reinhart, G. (2010). Telepresence in industrial applications: implementation issues for assembly tasks. *Presence: Teleoperators and Virtual Environments, 19* (5), pp. 415-429.

Reintsema, D., Landzettel, K., & Hirzinger, G. (2007). DLR's advanced telerobotic concepts and experiments for on-orbit servicing. In M. Ferre, M. Buss, R. Aracil, C. Melchiorri, & C. Balager (Eds.), *Advances in Telerobotics* (pp. 323-345). Berlin Heidelberg: Springer Verlag.

* Reiter, A., Nitsch, V., Reinhart, G., & Färber, B. (2008). Effects of Visual and Haptic Feedback on Telepresent Micro Assembly Tasks. *3rd International Conference on Changeable, Agile, Reconfigurable and Virtual Production (CARV).*

* Ren, J., Patel, R. V., McIsaac, K. A., Guiraudon, G., & Peters, T. M. (2008). Dynamic 3-D Virtual Fixtures for Minimally Invasive Beating Heart Procedures. *IEEE Transactions on Medical Imaging, 27* (8), pp. 1061-1070.

Reznick, R. K., & MacRae, H. (2006). Teaching surgical skills- changes in the wind. *The New England Journal of Medicine, 355* (25), pp. 2664-2669.

* Richard, P., Coiffet, P., Kheddar, A., & England, R. (1999). Human performance evaluation of two handle haptic devies in a dextrous virtual telemanipulation task. *IEEE/RSJ International Conference on Intelligent Robots and Systems*, (pp. 1543-1548).

Rosch, E. (1978). Principles of categorization. In E. Rosch, & B. Lloyd (Eds.), *Cognition and Categorization* (pp. 27-48). Hillsdale, NJ: Erlbaum.

Rosen, J., Hannaford, B., & Satava, R. M. (2010). *Surgical Robotics: Systems Applications and Visions* (1 ed.). Heidelberg Berlin: Springer Verlag.

Rosenberg, L. B. (1993). Virtual fixtures: Perceptual tools for telerobotic manipulation. *IEEE Annual International Symposium on Virtual Reality*, (pp. 76-82).

Salas, E., & Cannon-Bowers, J. (2001). The science of training: a decade of progress. *Annual Review of Psychology, 52*, pp. 471-479.

* Sallnäs, E. L. (2000). Improved precision in mediated collaborative manipulation of objects by haptic force feedback. In S. Brewster, & R. Murray-Smith (Eds.), *Haptic HCI 2000, Lecture Notes in Computer Science* (Vol. 2058, pp. 69-75). Springer-Verlag.

* Sallnäs, E. L., & Zhai, S. (2003). Collaboration meets Fitts' law: Passing virtual objects with and without haptic force feedback. *IFIP Conference on Human-Computer Interaction.*

Salvendy, G. (2006). *Handbook of human factors engineering* (3 ed.). New Jersey: John Wiley & Sons.

Sanders, A. F. (1980). Stage analysis of reaction processes. In G. Steimach, & J. Requin (Eds.), *Tutorials in Motor Behavior* (pp. 331-354). Amsterdam: North-Holland.

Scepkowski, L. A., & Cronin-Golomb, A. (2003). The alien hand: cases, categorizations, and anatomical correlates. *Behavioral and Cognitive Neuroscience Reviews, 2* (4), pp. 261-277.

Scheuchenpflug, R. (2001). Measuring presence in virtual environments. *Conference on Human Factors in Computing Systems*, (pp. 56-58).

* Schoonmaker, R. E., & Cao, C. G. (2006). Vibrotactile force feedback system for minimally invasive surgical procedures. *IEEE International Conference on Systems, Man, and Cybernetics*, (pp. 2464-2469).

Schwartz, S. P., Pomerantz, J. R., & Egeth, H. E. (1977). State and process limitations in information processing: An additive factors analysis. *Journal of Experimental Psychology: Human Perception and Performance, 3*, pp. 402-410.

Seymour, N. E., Gallagher, A. G., Roman, S. A., O'Brien, M. K., Bansal, V. K., Andersen, D. K., et al. (2002). Virtual reality training improves operating room performance. *Annals of Surgery, 236* (4), pp. 458-464.

Shen, X., Zhou, J., Hamam, A., Nourian, S., El-Far, N. R., Malric, F., et al. (2008). Haptic-enabled telementoring surgery simulation. *IEEE Multimedia, 15* (1), pp. 64-75.

Solis, J., Avizzano, C. A., & Bergamasco, M. (2002). Teaching to write japanese characters using a haptic interface. *10th Symposium On Haptic Interfaces For Virtual Environments & Teleoperator Systems (HAPTICS)*.

* Solis, J., Avizzano, C. A., & Bergamasco, M. (2003). Validating a skill transfer system based on reactive robots technology. *IEEE International Workshop on Robot and Human Interactive Communication*, (pp. 175-180).

* Srimathveeralli, G., & Thenkurussi, K. (2005). Motor skill training assistance using haptic attributes. *First Joint Eurohaptics Conference and Symposiuim on Haptic Interfaces for Virtual Environment and Teleoperator Systems*.

Stanney, K. M., Mollaghasemi, M., Reeves, L., Breaux, R., & Graeber, D. (2003). Usability engineering of virtual environments (VEs): identifying multiple criteria that drive effective VE system design. *International Journal of Human-Computer Studies, 58* (4), pp. 447-481.

* Steele, M., & Gillespie, R. B. (2001). Shared control between human and machine: using a haptic steering wheel to aid in land vehicle guidance. *Human Factors and Ergonomics Society Annual Meeting Proceedings* (5), pp. 1671-1675.

Stredney, D., Wiet, G. J., Yagel, R., Sessanna, D., Kurzion, Y., Fontana, M., et al. (1998). A comparative analysis of integrating visual representations with haptic displays. *Studies in Health Technology and Informatics, 50*, pp. 20-26.

* Ström, P., Hedman, L., Särna, L., Kjellin, A., Wredmark, T., & Felländer-Tsai, L. (2006). Early exposure to haptic feedback enhances performance in surgical simulator training: a prospective randomized crossover study in surgical residents. *Surgical Endoscopy*, pp. 1383-1388.

Stylopoulos, N., & Rattner, D. (2003). Robotics and ergonomics. *The surgical clinics of North America , 83* (6), pp. 1321-1337.

Sutherland, I. (1965). The ultimate display. *International Federation of Information Processing*, (pp. 506-508).

Syrdal, D. S., Koay, K. L., Walters, M. L., & Dautenhahn, K. (2007). A personalized robot companion? The role of individual differences on spatial preferences in HRI scenarios. *16th IEEE International Conference on Robot & Human Interactive Communication*, (pp. 1143-1148).

Szalma, J. L., Warm, J. S., Matthews, G., Dember, W. N., Weiler, E. M., Meier, A., et al. (2004). Effects of sensory modality and task duration on performance, workload, and stress in sustained attention. *Human Factors, 46* (2), pp. 219-233.

Takesue, N., Murayama, H., Fujiwara, K., Matsumoto, K., Konosu, H., & Fujimoto, H. (2006). Kinesthetic assistance for improving task performance- The case of window installation assist. *International Journal of Automation Technology, 3* (6), pp. 663-670.

Tan, H. Z. (2000). Haptic Interfaces. *Communications of the ACM, 43* (3), pp. 40-41.

Tan, H. Z., Eberman, B., Srinivasan, M. A., & Cheng, B. (1994). Human factors for the design of force-reflecting haptic interfaces. In C. Radcliffe (Ed.), *Dynamic Systems and Control* (Vol. 55(1), pp. 353-359). The American Society of Mechanical Engineers.

Tavakoli, M., Patel, R. V., & Moallem, M. (2005). Haptic feedback and sensory substitution during telemanipulated suturing. *First Joint Eurohaptics Conference and Symposium on Haptic Interfaces for Virtual Environment and Teleoperator Systems*, (pp. 543-544).

Teo, C. L., Burdet, E., & Lim, H. P. (2002). A robotic teacher of Chinese handwriting. *10th Symposium on Haptic Interfaces for Virtual Environment and Teleoperator Systems (HAPTICS)*.

Thurstone, L. L. (1927). The method of paired comparisons for social values. *Journal of Abnormal and Social Psychology, 21*, pp. 384-400.

Tsumaki, Y., Naruse, H., Nenchev, D. N., & Uchiyama, M. (1998). Design of a compact 6-DOF haptic interface. *IEEE International Conference on Robotics & Automation*, (pp. 2580-2585).

Turro, N., Khatib, O., & Coste-Maniere, E. (2001). Haptically augmented teleoperation. *IEEE International Conference on Robotics and Automation*, (pp. 386-392).

Ueberle, M. W. (2006). *Design, Control, and Evaluation of a Family of Kinesthetic Haptic Interfaces*. Dissertation, Technische Universität München.

Ueberle, M., Mock, N., & Buss, M. (2007). Design, control, and evaluation of a hyper-redundant haptic device. In M. Ferre, M. Buss, R. Aracil, C. Melchiorri, & C. Balaguer (Eds.), *Advances in Telerobotics* (pp. 25-44). Berlin Heidelberg: Springer Verlag.

* Unger, B. J., Berkelman, P. M., Thompson, A., Lederman, S., Klatzky, R. L., & Hollis, R. L. (2002). Virtual peg-in-hole performance using a 6-DOF magnetic levitation haptic device: Comparison with real forces and with visual guidance alone. *10th Symposium on Haptic Interfaces for Virtual Environment and Teleoperator Systems*, (pp. 263-270).

Utsumi, M., Hirabayashi, T., & Yoshie, M. (2002). Development for teleoperation underwater grasping system in unclear environment. *International Symposium on Underwater Technology*, (pp. 349-353).

Valeriani, M., Ranhi, F., & Giaquinto, S. (2003). The effects of aging on selective attention to touch: A reduced inhibitory control in elderly subjects? *International Journal of Psychophysiology* (49), pp. 75-87.

Van der Linde, R. Q., Lammertse, P., Frederiksen, E., & Ruiter, B. (2002). The HapticMaster, a new high-performance haptic interface. *EuroHaptics*, (pp. 1-5).

* Viau, A., Najm, M., & Chapman, C. E. (2005). Effect of tactile feedback on movement speed and precision during work-related tasks using a computer mouse. *Human Factors: The Journal of the Human Factors and Ergonomics Society* (47), pp. 816-826.

* Wagner, C. R., Howe, R. D., & Stylopoulos, N. (2002). The role of force feedback in surgery: analysis of blunt dissection. *10th Symposium on Haptic Interfaces for Virtual Environment and Teleoperator Systems*.

* Wall, S. A., & Harwin, W. S. (2000). Quantification of the effects of haptic feedback during a motor skills task in a simulated environment. *Proceedings of the 2nd PHANToM Users Research Symposium*, (pp. 61-69).

Ware, C., & Balakrishnan, R. (1994). Reaching for objects in VR displays: lag and frame rate. *Computer-Human Interaction , 1* (4), pp. 331-356.

Wickens, A. (2000). *Foundations of Biopsychology.* UK: Pearson/Prentice Hall.

Wickens, C. D. (1984). Processing resources in attention. In R. Parasuraman, & D. Davies (Eds.), *Varieties of Attention.* London: Academic Press.

Wickens, C. D., Lee, J. D., Liu, Y., & Gordon Becker, S. E. (2004). *An introduction to human factors engineering* (2 ed.). New Jersey: Pearson Prentice Hall.

Williams, R. L., Chen, M. Y., & Seaton, J. M. (2002). Haptics-Augmented High School Physics Tutorials. *International Journal of Virtual Reality, 5* (1).

Williams, R. L., Srivastave, M., Conatser, R. R., & Howell, J. N. (2004). Implementation and evaluation of a haptic playback system. *Haptics-e, 3* (3).

Yang, X., Chen, Q., Petriu, D. C., & Petriu, E. M. (2004). Internet-based teleopeartion of a robot manipulator for education. *3rd IEEE International Workshop on Haptic, Audio and Visual Environments and Their Applications*, (pp. 7-11).

* Yang, X.-D., Bischof, W. F., & Boulanger, P. (2008). Validating the performance of haptic motor skill training. *IEEE Symposium on Haptic Interfaces for Virtual Environments and Teleoperator Systems*, (pp. 129-135).

Yokokohji, Y., Hollis, R. L., & Kanade, T. (1996). Toward machine mediated training of motor skills. *IEEE International Workshop on Robot and Human Communication*, (pp. 32-37).

Zaeh, M. F., & Reiter, A. (2006). Precise positioning in a telepresent micro-assembly system. *IEEE International Workshop on Haptic Audio Visual Environments and their Applications*, (pp. 44-47).

Zhai, S., & Senders, J. W. (1997). Investigating coordination in multidegree of freedom control II: correlation analysis in 6 DOF tracking. *41st Annual Meeting of the Human Factors and Ergonomic Society*, (pp. 1254-1258).

Appendix A.

Assumptions for the applicability of parametric statistical tests

1. Interval Data:

Data should be measured at least at the interval level.

2. Independence:

Data from different participants should be independent.

3. Homogeneity of variance:

The variance of one variable should be stable at all levels of the other variable. Levene's tests indicate whether the assumption of homogeneity of variance had been violated. However, with a large sample size of 800 cases, very small differences in group variances can produce a significant Levene's test (Field, 2009, p. 133). In this case, variance ratios are also inspected for an indication of heterogeneity of variance.

4. Normal distribution:

The data should be from one or more normally distributed populations. Skewness and kurtosis values (transformed into z-scores) can be used as indicators. Significant skewness and kurtosis values can be expected with large sample sizes, as the high number of cases results in small standard errors. Kolmogorov-Smirnov tests and histograms are also used to check the data for this assumption (Field, 2009). In cases where a non-normal distribution can be precluded on logical grounds, and the samples size is $N \geq 30$, the central limit theorem applies (e.g. Field (2009), Bortz & Döring (2006)). In these cases, a close to normal distribution can be assumed.

Appendix B.

Test statistics on task facilitative effects of force feedback (Ch. II)

Table 8. Mean values and deviations of average force standard deviations (N) for each trial.

Trial no.	Force Deviation (N)	
	M	SD
1st trial	109.92	55.63
2nd trial	170.47	57.40
3rd trial	154.72	62.56

Table 9. Mean values and standard deviations of mean force values (N) for each trial conducted with and without force feedback.

Type of Feedback	Trial no.	Mean Force (N)	
		M	SD
With Force Feedback	T1	420.39	77.74
	T2	301.77	125.88
	T3	301.99	129.56
Without Force Feedback	T1	385.21	61.63
	T2	277.87	96.26
	T3	342.63	111.63

Table 10. Pearson correlation coefficients (r) and probability values (p) for force measures (N) task completion time (sec.) by feedback type.

	Mean Force		Max. Force		Force Deviation	
	r	p*	r	p*	r	p*
Task completion time without force feedback	-.06	.73	-.09	.60	-.03	.86
Task completion time with force feedback	-.30	.08	-.31	.08	.12	.49

*accepted α-level $p = <.05$

Table 11. Mean values and deviations of task completion times (sec.) for each trial.

	Task Completion Time (sec.)	
Trial no.	M	SD
1st trial	8.98	5.67
2nd trial	13.04	7.36
3rd trial	12.50	7.94

Appendix C.

Test statistics on task facilitative effects of haptic expert demonstrations (Ch. III)

Table 12. Results of one-sample t-tests comparing mean negative deviations to the task completion time of the expert.

Group	Trial No.	t-Statistic	α-Probability*
No Haptic demonstrations	T1	t(15)=-5.72	<.001
	T2	t(14)=-3.83	<.005
	T3	t(15)=-3.56	<.005
	T4	t(15)=-2.79	=.014
	T5	t(15)=-3.15	=.007
	T6	t(15)=-2.27	=.039
	T7	t(15)=-1.34	=.200
	T8	t(15)=-1.78	=.095
Haptic demonstrations	T1	t(15)= -5.38	<.001
	T2	t(15)= -3.59	<.005
	T3	t(15)=-3.60	<.005
	T4	t(15)=-2.58	=.021
	T5	t(15)=-2.75	=.015
	T6	t(14)=-1.69	=.114
	T7	t(14)=-1.29	=.217
	T8	t(15)=-1.06	=.308

*accepted α-level = $p < .006$

Table 13. Means, medians, and standard deviations for successful pick-and-place attempts (percentage).

Task	Track	Instruction	Median	Mean	Std. Dev.
Picking	Demonstrated	No Haptic demonstrations	100%	96.88%	5.59%
		Haptic demonstrations	93.75%	90.63%	11.64%
	Novel	No Haptic demonstrations	100%	99.22%	3.13%
		Haptic demonstrations	100%	96.09%	8.80%
Placing	Demonstrated	No Haptic demonstrations	71.43%	67.86%	16.71%
		Haptic demonstrations	57.14%	60.30%	22.10%
	Novel	No Haptic demonstrations	66.96%	65.02%	16.98%
		Haptic demonstrations	64.58%	63.37%	16.17%

Appendix D.

Test statistics on task facilitative effects of haptic assistance functions (Ch. III)

Table 14. T-tests and probability values with effect sizes for each comparison of control ratings between different types of assistance.

Comparison	t-Statistic	α-Probability*	Effect Size
VFC-AVF	t(30) = 6.86	p <.001	r = .78
VFC-VD	t(30) = - 6.23	p <.001	r = .75
VFC-NA	t(30) = - 5.56	p <.001	r = .71
AVF-VD	t(30) = -16.62	p <.001	r = .95
AVF-NA	t(30) = -15.27	p <.001	r = .94
VD-NA	t(30) = 1.43	p = .16	

*accepted α-level p = <.008

Table 15. T-tests and probability values with effect sizes for each comparison of the collision times between different types of assistance.

Comparison	t-Statistic	α-Probability*	Effect Size
VFC-AVF	t(30) = -15.75	p <.001	r = .94
VFC-VD	t(30) = -7.60	p <.001	r = .81
VFC-NA	t(30) = 7.61	p <.001	r = .81
AVF-VD	t(30) = -1.57	p = .13	
AVF-NA	t(30) = 2.54	p = .02	
VD-NA	t(30) = 1.70	p = .10	

*accepted α-level p = <.008

Table 16. T-tests and probability values with effect sizes for each comparison of the task completion times between different types of assistance.

Comparison	t-Statistic	α-Probability*	Effect Size
VFC-AVF	t(30) = -2.90	p <.008	r = .47
VFC-VD	t(30) = -10.17	p <.001	r = .88
VFC-NA	t(30) = 9.04	p <.001	r = .86
AVF-VD	t(30) = -4.70	p <.001	r = .65
AVF-NA	t(30) = 4.07	p <.001	r = .60
VD-NA	t(30) = -3.68	p < .001	r = .56

*accepted α-level p = <.008

Appendix E.

Meta-analysis results statistics for the effect of vibrotactile feedback on task performance[17] (Ch. IV)

1. Overall performance

Stratum	* Difference	SE	Approximate 95% CI		% Weight (fixed, random)		Study
1	0	0.500009	-0.98	0.98	8.26569	10.029412	3
2	1.458628	0.461743	0.58	2.39	9.692462	11.614911	4
3	1.150227	0.727054	-0.2	2.65	3.909323	4.932428	5
4	0.692539	0.653073	-0.55	2.01	4.845196	6.061361	9
5	0.890407	0.428579	0.09	1.77	11.250522	13.302025	16
6	0.584892	0.188779	0.22	0.96	57.986621	48.956279	17
7	1.413538	0.714299	0.09	2.89	4.050188	5.103586	23

Non-combinability of studies
Cochran Q = 6.596763 (df = 6) P = 0.3598
Moment-based estimate of between studies variance = 0.019595
I² (inconsistency) = 9% (95% CI = 0% to 62.1%)

Random effects (DerSimonian-Laird)
Pooled * difference = 0.745054 (95% CI = 0.422763 to 1.067346)
Z (test test * difference differs from 0) = 4.530926 P < 0.0001

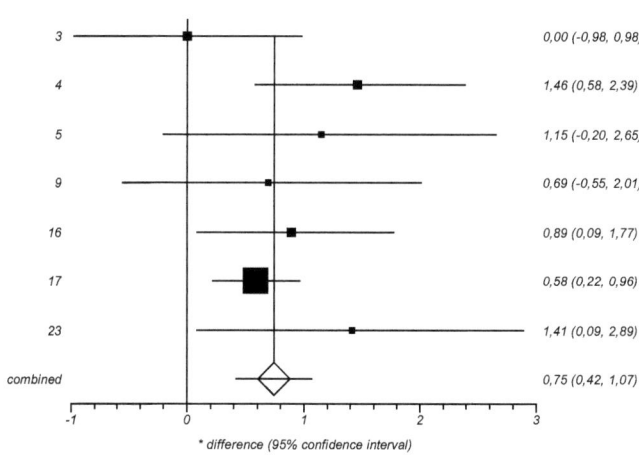

---

[17] All tables and figures in the appendices that pertain to meta-analysis were produced with StatsDirect software (Version 2.7.8) by StatsDirect Ltd.

2. Force application

Stratum	* Difference	SE	Approximate 95% CI		% Weight (fixed, random)		Study
1	0	0.500009	-0.98	0.98	22.729779	20.935885	3
2	-1.742522	0.482152	-2.73	-0.84	24.444646	21.155318	4
3	1.150227	0.727054	-0.2	2.65	10.750227	17.974384	5
4	0.890407	0.428579	0.09	1.77	30.937755	21.790662	16
5	1.413538	0.714299	0.09	2.89	11.137592	18.143752	23

Non-combinability of studies
Cochran Q = 23.496502 (df = 4) P = 0.0001
Moment-based estimate of between studies variance = 1.440903
I² (inconsistency) = 83% (95% CI = 51.8% to 91%)

Random effects (DerSimonian-Laird)
Pooled * difference = 0.288605 (95% CI = -0.877545 to 1.454754)
Z (test test * difference differs from 0) = 0.485062 P = 0.6276

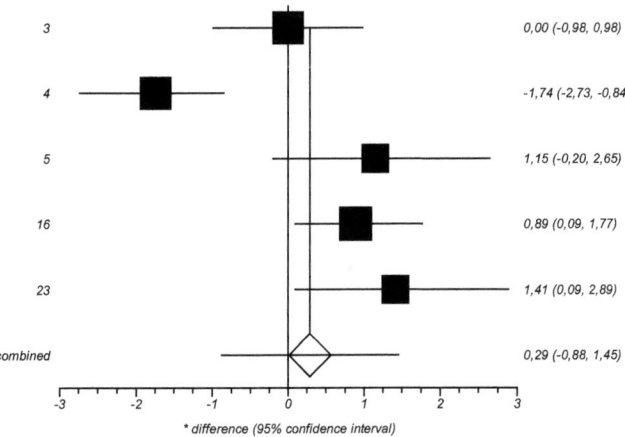

3. Task completion time

Stratum	* Difference	SE	Approximate 95% CI		% Weight (fixed, random)		Study
1	1.458628	0.461743	0.58	2.39	13.424321	26.401722	4
2	0.386552	0.676033	-0.91	1.74	6.262644	15.053519	5
3	0.584892	0.188779	0.22	0.96	80.313035	58.544759	17

Non-combinability of studies
Cochran Q = 3.282451 (df = 2) P = 0.1937
Moment-based estimate of between studies variance = 0.110215
I^2 (inconsistency) = 39.1% (95% CI = 0% to 82.2%)

Random effects (DerSimonian-Laird)
Pooled * difference = 0.785716 (95% CI = 0.212988 to 1.358444)
Z (test test * difference differs from 0) = 2.68884 P = 0.0072

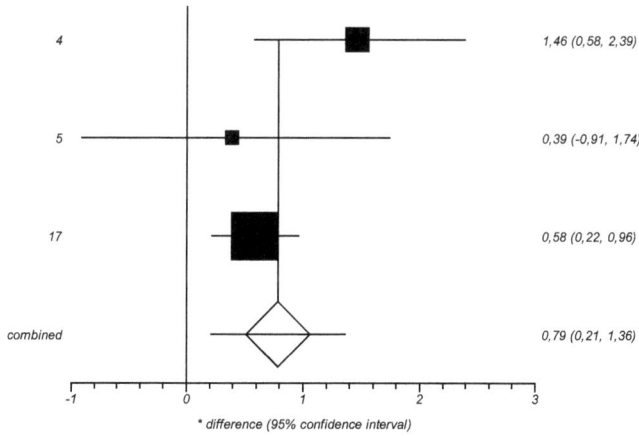

4. Error rate

Stratum	* Difference	SE	Approximate 95% CI		% Weight (fixed, random)		Study
1	0.692539	0.653073	-0.55	2.01	23.021988	23.021988	9
2	0.43307	0.357149	-0.26	1.14	76.978012	76.978012	23

Non-combinability of studies
Cochran Q = 0.12151 (df = 1) P = 0.7274
Moment-based estimate of between studies variance = 0
I² (inconsistency) = *% (95% CI = *% to *%)

Random effects (DerSimonian-Laird)
Pooled * difference = 0.492805 (95% CI = -0.121355 to 1.106965)
Z (test test * difference differs from 0) = 1.572686 P = 0.1158

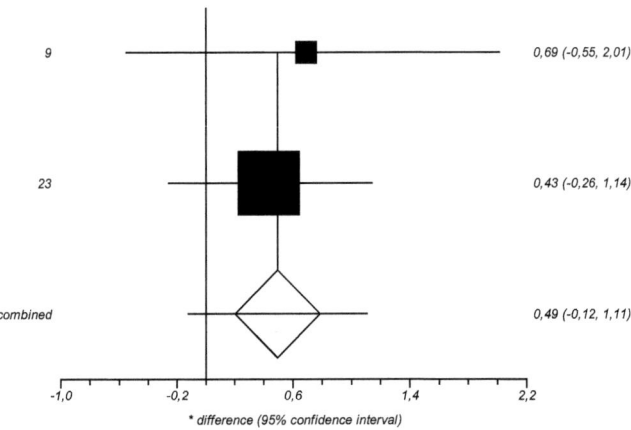

Appendix F.

Meta-analysis results statistics for the effect of force feedback on task performance (Ch. IV)

1. Overall performance

Stratum	*Difference	SE	Approximate 95% CI		% Weight (fixed, random)		Study
1	2.248199	0.714299	0.93	3.73	1.297248	2.750165	1
2	1.48462	0.290822	0.94	2.08	7.825813	4.677742	2
3	0.068586	0.400518	-0.71	0.86	4.126101	4.158444	6
4	0.554136	0.293373	-0.02	1.13	7.690304	4.666291	7
5	4.724189	1.030631	2.87	6.91	0.623127	1.792364	8
6	0.974679	0.336741	0.34	1.66	5.837022	4.465829	11
7	0.852422	0.418375	0.05	1.69	3.78139	4.071291	13
8	0.087721	0.316332	-0.53	0.71	6.614482	4.561404	14
9	1.850158	0.64542	0.65	3.18	1.588906	3.024385	15
10	2.974618	1.063795	1.1	5.27	0.58488	1.717111	18
11	0.237171	0.635216	-1	1.49	1.640365	3.067056	26
12	0.548209	0.293373	-0.01	1.14	7.690304	4.666291	30
13	1.508494	0.76532	0.1	3.1	1.130047	2.562858	33
14	2.765601	0.760218	1.38	4.36	1.145267	2.580982	35
15	-0.871261	0.428579	-1.73	-0.05	3.603468	4.02148	36
16	2.86417	0.507662	1.91	3.9	2.568225	3.639956	37
17	0.95074	0.512765	-0.03	1.98	2.51737	3.615834	38
18	1.064939	0.456641	0.19	1.98	3.174192	3.884866	39
19	3.403861	0.969406	1.67	5.47	0.704323	1.942441	40
20	0.540749	0.456641	-0.32	1.47	3.174192	3.884866	41
21	0.474342	0.45409	-0.39	1.39	3.209957	3.897248	42
22	1.064368	0.686237	-0.16	2.53	1.405512	2.858943	43
23	0.680861	0.33419	0.04	1.35	5.926477	4.477878	45
24	1.954288	0.783178	0.53	3.6	1.079102	2.500478	46
25	1.861638	0.380109	1.17	2.66	4.581067	4.25774	47
26	0.249971	0.535724	-0.78	1.32	2.306219	3.508335	49
27	0.849033	0.459192	-0.03	1.77	3.139021	3.872493	50
28	0.522148	0.244902	0.05	1.01	11.03562	4.875228	53

Non-combinability of studies
Cochran Q = 101.614038 (df = 27) P < 0.0001
Moment-based estimate of between studies variance = 0.522711
I^2 (inconsistency) = 73.4% (95% CI = 60.1% to 80.9%)

Random effects (DerSimonian-Laird)
Pooled * difference = 1.083552 (95% CI = 0.75321 to 1.413894)
Z (test test * difference differs from 0) = 6.428863 P < 0.0001

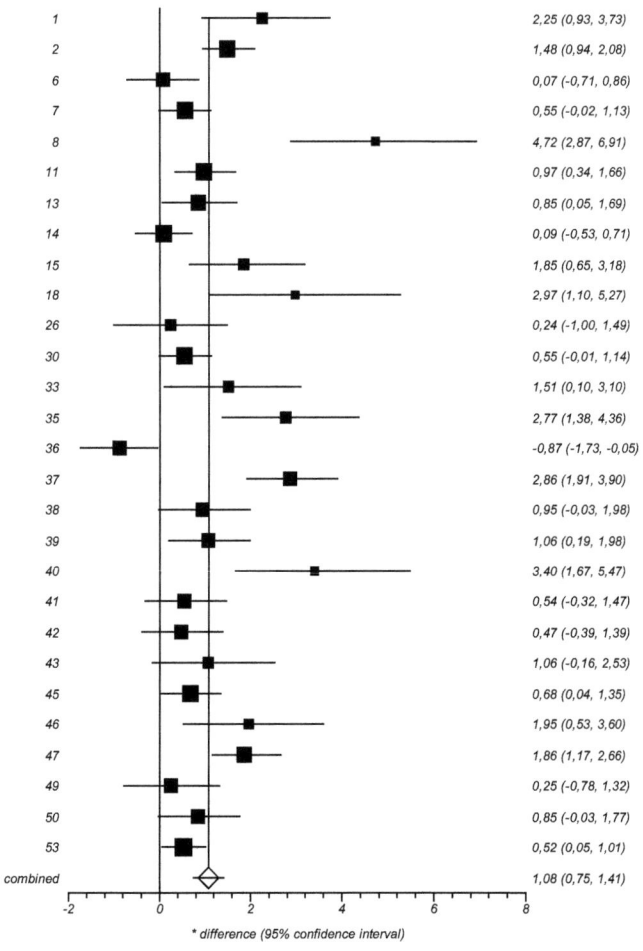

2. Force application

Stratum	* Difference	SE	Approximate 95% CI		% Weight (fixed, random)		Study
1	4.724189	1.030631	2.87	6.91	2.308778	6.339557	8
2	0.852422	0.418375	0.05	1.69	14.010616	15.217755	13
3	2.974618	1.063795	1.1	5.27	2.167069	6.062632	18
4	0.407934	0.45409	-0.46	1.32	11.89337	14.504315	41
5	0.474342	0.45409	-0.39	1.39	11.89337	14.504315	42
6	1.064368	0.686237	-0.16	2.53	5.207633	10.372878	43
7	0.849033	0.459192	-0.03	1.77	11.630541	14.403336	50
8	0.522148	0.244902	0.05	1.01	40.888622	18.595211	53

Non-combinability of studies
Cochran Q = 21.350457 (df = 7) P = 0.0033
Moment-based estimate of between studies variance = 0.458449
I² (inconsistency) = 67.2% (95% CI = 8.5% to 82.7%)

Random effects (DerSimonian-Laird)
Pooled * difference = 1.06731 (95% CI = 0.458765 to 1.675854)
Z (test test * difference differs from 0) = 3.437529 P = 0.0006

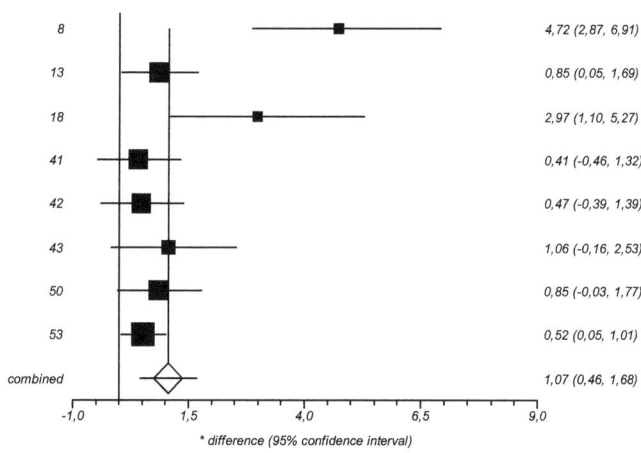

Summary meta-analysis plot [random effects]

* difference (95% confidence interval)

3. Task completion time

Stratum	* Difference	SE	Approximate 95% CI		% Weight (fixed. random)		Study
1	2.248199	0.714299	0.93	3.73	1.588835	3.513953	1
2	1.48462	0.290822	0.94	2.08	9.584847	6.222659	2
3	0.068586	0.400518	-0.71	0.86	5.053539	5.471235	6
4	0.554136	0.293373	-0.02	1.13	9.418879	6.205909	7
5	1.323546	0.57399	0.25	2.5	2.460537	4.310842	8
6	0.087721	0.316332	-0.53	0.71	8.101241	6.052873	11
7	0.774039	0.415824	-0.03	1.6	4.688346	5.364502	13
8	0.087721	0.316332	-0.53	0.71	8.101241	6.052873	14
9	0.237171	0.635216	-1	1.49	2.009075	3.944467	26
10	1.508494	0.76532	0.1	3.1	1.384052	3.262106	33
11	-0.920208	0.43113	-1.79	-0.1	4.361356	5.25808	36
12	2.86417	0.507662	1.91	3.9	3.145493	4.737237	37
13	0.95074	0.512765	-0.03	1.98	3.083208	4.703475	38
14	0.752296	0.441335	-0.1	1.63	4.162006	5.187407	39
15	3.403861	0.969406	1.67	5.47	0.862636	2.441493	40
16	0.540749	0.456641	-0.32	1.47	3.887665	5.081954	41
17	0.085381	0.448988	-0.79	0.97	4.021329	5.134589	42
18	0.311897	0.316332	-0.3	0.94	8.101241	6.052873	47
19	-0.481426	0.543377	-1.59	0.54	2.74559	4.504134	49
20	0.522148	0.247454	0.05	1.02	13.238886	6.49734	53

Non-combinability of studies
Cochran Q = 72.585972 (df = 19) P < 0.0001
Moment-based estimate of between studies variance = 0.467605
I^2 (inconsistency) = 73.8% (95% CI = 57.1% to 82.1%)

Random effects (DerSimonian-Laird)
Pooled * difference = 0.694417 (95% CI = 0.331107 to 1.057727)
Z (test test * difference differs from 0) = 3.746204 P = 0.0002

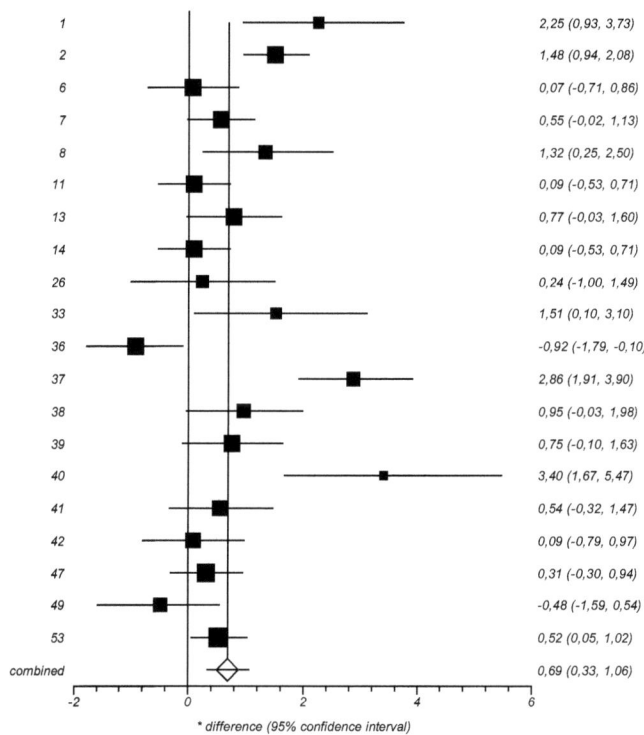

4. Error Rate

Stratum	* Difference	SE	Approximate 95% CI		% Weight (fixed, random)		Study
1	1.152923	0.602562	0.018	2.38	3.756339	7.059636	1
2	0.068823	0.260209	-0.44	0.58	20.143001	10.601929	2
3	0.974679	0.336741	0.34	1.66	12.027536	9.83838	11
4	1.850158	0.64542	0.65	3.18	3.274036	6.656882	15
5	0.548209	0.293373	-0.01	1.14	15.846335	10.281283	30
6	2.765601	0.760218	1.38	4.36	2.359891	5.679362	35
7	-0.871261	0.428579	-1.73	-0.05	7.425162	8.857025	36
8	1.064939	0.456641	0.19	1.98	6.540613	8.555417	39
9	0.680861	0.33419	0.04	1.35	12.211863	9.864998	45
10	1.954288	0.783178	0.53	3.6	2.223554	5.501941	46
11	1.861638	0.380109	1.17	2.66	9.439565	9.378744	47
12	0.249971	0.535724	-0.78	1.32	4.752104	7.724403	49

Non-combinability of studies
Cochran Q = 44.63049 (df = 11) P < 0.0001
Moment-based estimate of between studies variance = 0.520956
I² (inconsistency) = 75.4% (95% CI = 52.1% to 84.7%)

Random effects (DerSimonian-Laird)
Pooled * difference = 0.903716 (95% CI = 0.414078 to 1.393353)
Z (test test * difference differs from 0) = 3.617474 P = 0.0003

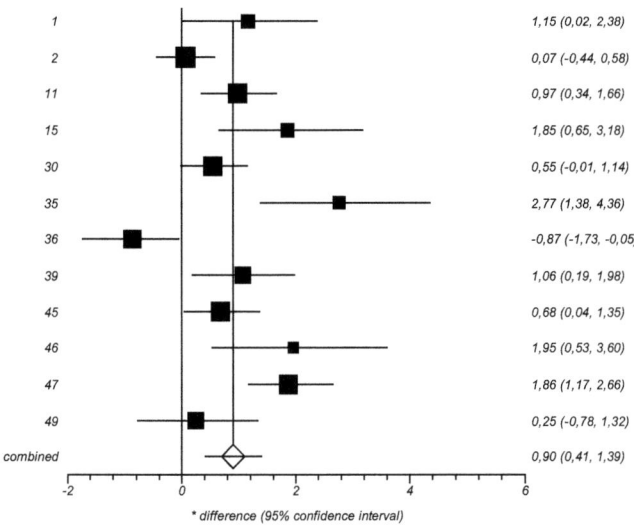

Summary meta-analysis plot [random effects]

Appendix G.

Meta-analysis results statistics for the effect of haptic expert demonstrations on task performance (Ch. IV)

1. Overall performance

Stratum	* Difference	SE	Approximate 95% CI		% Weight (fixed, random)		Study
1	-0.02958	0.331639	-0.68	0.62	26.979402	19.27535	27
2	-0.038987	0.448988	-0.92	0.84	14.719521	15.512247	28
3	0.00383	0.579092	-1.13	1.14	8.848452	12.071201	29
4	0.955186	0.471947	0.06	1.91	13.322188	14.843369	31
5	1.935314	0.780627	0.51	3.57	4.869408	8.299444	32
6	0.71807	0.59695	-0.42	1.92	8.326976	11.664925	34
7	-0.551181	0.3597	-1.27	0.14	22.934052	18.333464	54

Non-combinability of studies
Cochran Q = 13.378437 (df = 6) P = 0.0374
Moment-based estimate of between studies variance = 0.267633
I^2 (inconsistency) = 55.2% (95% CI = 0% to 78.8%)

Random effects (DerSimonian-Laird)
Pooled * difference = 0.273827 (95% CI = -0.254953 to 0.802607)
Z (test test * difference differs from 0) = 1.01496 P = 0.3101

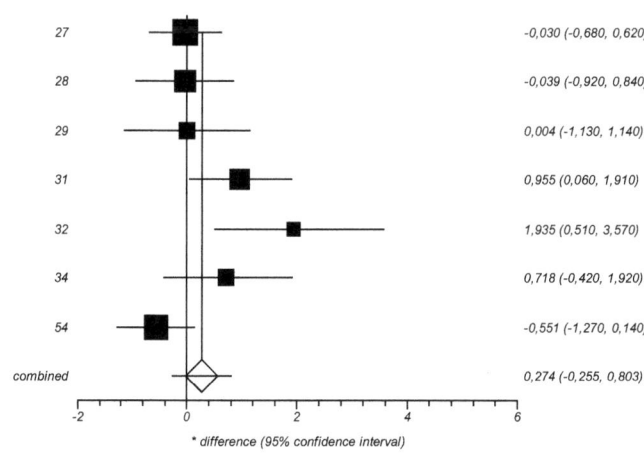

Summary meta-analysis plot [random effects]

2. Force application

Stratum	*Difference	SE	Approximate 95% CI		% Weight (fixed, random)		Study
1	0.00383	0.579092	-1.13	1.14	64.503152	53.672703	29
2	1.935314	0.780627	0.51	3.57	35.496848	46.327297	32

Non-combinability of studies
Cochran Q = 3.948904 (df = 1) P = 0.0469
Moment-based estimate of between studies variance = 1.392953
I² (inconsistency) = *% (95% CI = *% to *%)

Random effects (DerSimonian-Laird)
Pooled * difference = 0.898634 (95% CI = -0.989072 to 2.786341)
Z (test test * difference differs from 0) = 0.933032 P = 0.3508

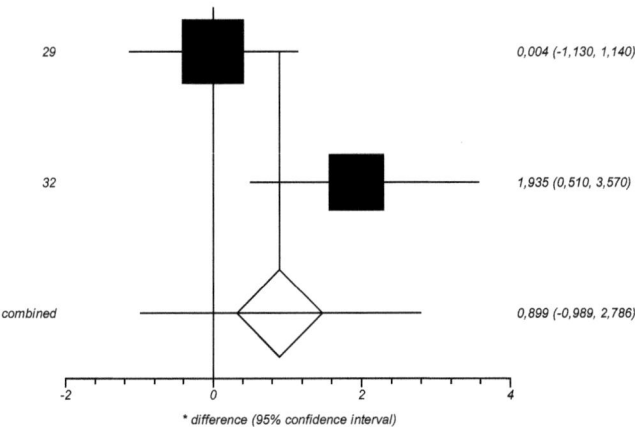

3. Task completion time

Stratum	* Difference	SE	Approximate 95% CI		% Weight (fixed. random)		Study
1	0.955186	0.471947	0.06	1.91	36.744538	47.943054	31
2	-0.551181	0.3597	-1.27	0.14	63.255462	52.056946	54

Non-combinability of studies
Cochran Q = 6.444244 (df = 1) P = 0.0111
Moment-based estimate of between studies variance = 0.95851
I^2 (inconsistency) = *% (95% CI = *% to *%)

Random effects (DerSimonian-Laird)
Pooled * difference = 0.171018 (95% CI = -1.303945 to 1.64598)
Z (test test * difference differs from 0) = 0.227252 P = 0.8202

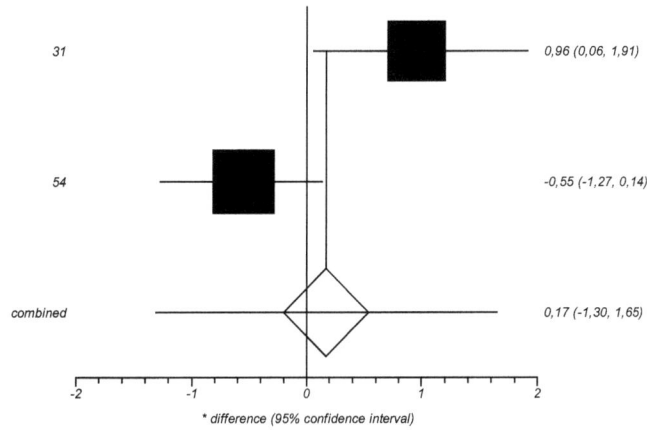

Summary meta-analysis plot [random effects]

1. Error rate

Stratum	* Difference	SE	Approximate 95% CI		% Weight (fixed. random)		Study
1	-0.02958	0.331639	-0.68	0.62	53.930869	53.930869	27
2	-0.038987	0.448988	-0.92	0.84	29.423802	29.423802	28
3	0.71807	0.59695	-0.42	1.92	16.64533	16.64533	34

Non-combinability of studies
Cochran Q = 1.319455 (df = 2) P = 0.517
Moment-based estimate of between studies variance = 0
I^2 (inconsistency) = 0% (95% CI = 0% to 72.9%)

Random effects (DerSimonian-Laird)
Pooled * difference = 0.092101 (95% CI = -0.385244 to 0.569445)
Z (test test * difference differs from 0) = 0.378163 P = 0.7053

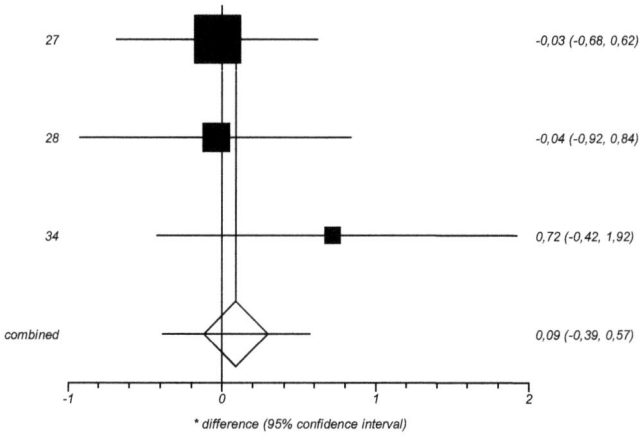

Appendix H.

Meta-analysis results statistics for the effect of haptic assistance on task performance[18] (Ch. IV)

1. Overall performance

Stratum	*Difference	SE	Approximate 95% CI		% Weight (fixed. random)		Study
1	1.271236	0.494907	0.37	2.31	9.700478	11.933809	10
2	2.5215	0.704095	1.23	3.99	4.79268	8.392913	19
3	1.516047	0.538275	0.5	2.61	8.200337	11.097409	20
4	2.091411	0.443886	1.29	3.03	12.058634	12.979556	21
5	1.908453	0.308679	1.34	2.55	24.935947	15.948377	22
6	3.228882	1.449006	0.77	6.45	1.131618	2.896754	24
7	1.094249	0.323986	0.49	1.76	22.635452	15.6086	25
8	2.618109	0.71685	1.3	4.11	4.62364	8.216737	44
9	0.155949	0.446437	-1.03	0.72	11.921215	12.925846	52

Non-combinability of studies
Cochran Q = 20.316578 (df = 8) P = 0.0092
Moment-based estimate of between studies variance = 0.349571
I² (inconsistency) = 60.6% (95% CI = 0% to 79.3%)

Random effects (DerSimonian-Laird)
Pooled * difference = 1.60701 (95% CI = 1.084956 to 2.129063)
Z (test test * difference differs from 0) = 6.033252 P < 0.0001

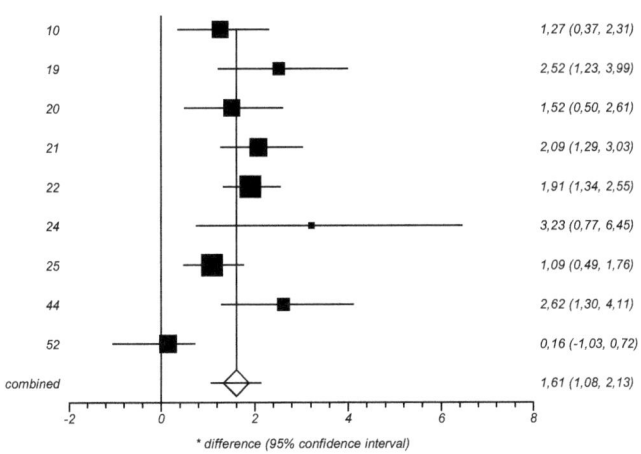

---

[18] Tables and figures were produced with StatsDirect software (Version 2.7.8) by StatsDirect Ltd.

2. Force application

g = 1.27, p<.0001   Approximate 95% CI: 0.37 – 2.31   Study 10

3. Task completion time

Stratum	*Difference	SE	Approximate 95% CI		% Weight (fixed, random)		Study
1	0.493315	0.45409	-0.37	1.41	13.275046	16.260871	10
2	1.429816	0.584194	0.34	2.63	8.020567	12.675169	19
3	1.14253	0.38266	0.43	1.93	18.693625	18.587387	21
4	0.727967	0.26276	0.23	1.26	39.646202	22.813494	22
5	3.228882	1.449006	0.77	6.45	1.303705	3.275003	24
6	2.618109	0.71685	1.3	4.11	5.326763	9.889079	44
7	0.155949	0.446437	-1.03	0.72	13.734092	16.498998	52

Non-combinability of studies
Cochran Q = 13.552386 (df = 6) P = 0.0351
Moment-based estimate of between studies variance = 0.271318
I² (inconsistency) = 55.7% (95% CI = 0% to 79.1%)

Random effects (DerSimonian-Laird)
Pooled * difference = 1.030273 (95% CI = 0.48412 to 1.576425)
Z (test test * difference differs from 0) = 3.697314 P = 0.0002

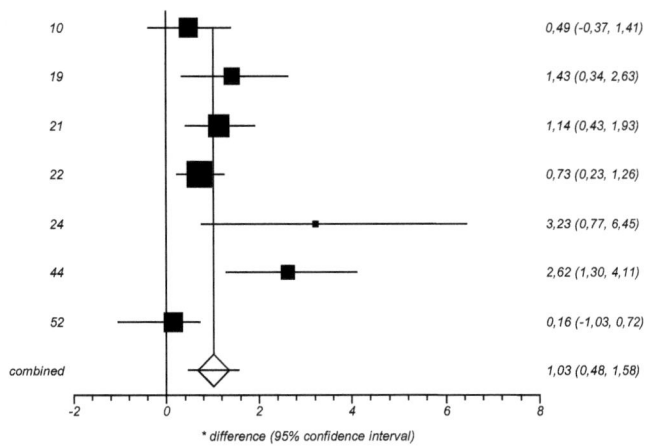

4. Error rate

Stratum	* Difference	SE	Approximate 95% CI		% Weight (fixed. random)		Study
1	2.5215	0.704095	1.23	3.99	6.012787	10.589432	19
2	1.516047	0.538275	0.5	2.61	10.287955	14.513978	20
3	2.091411	0.443886	1.29	3.03	15.128486	17.419087	21
4	1.908453	0.308679	1.34	2.55	31.284069	22.323305	22
5	0.228218	0.579092	-0.88	1.39	8.888782	13.413432	24
6	1.094249	0.323986	0.49	1.76	28.39792	21.740766	25

Non-combinability of studies
Cochran Q = 11.932476 (df = 5) P = 0.0357
Moment-based estimate of between studies variance = 0.266125
I^2 (inconsistency) = 58.1% (95% CI = 0% to 81%)

Random effects (DerSimonian-Laird)
Pooled * difference = 1.545896 (95% CI = 0.98919 to 2.102602)
Z (test test * difference differs from 0) = 5.442549 P < 0.0001

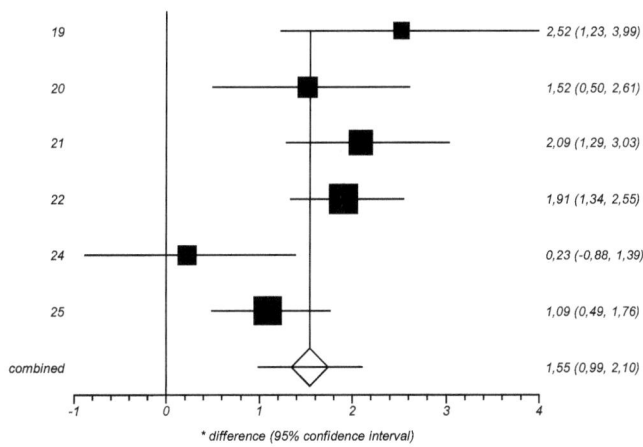

Summary meta-analysis plot [random effects]

i want morebooks!

Buy your books fast and straightforward online - at one of world's fastest growing online book stores! Environmentally sound due to Print-on-Demand technologies.

Buy your books online at
www.get-morebooks.com

Kaufen Sie Ihre Bücher schnell und unkompliziert online – auf einer der am schnellsten wachsenden Buchhandelsplattformen weltweit! Dank Print-On-Demand umwelt- und ressourcenschonend produziert.

Bücher schneller online kaufen
www.morebooks.de

 VDM Verlagsservicegesellschaft mbH
Heinrich-Böcking-Str. 6-8   Telefon: +49 681 3720 174   info@vdm-vsg.de
D - 66121 Saarbrücken   Telefax: +49 681 3720 1749   www.vdm-vsg.de

Printed by Books on Demand GmbH, Norderstedt / Germany